MULTIMEDIA DATABASE MANAGEMENT SYSTEMS

MULTIMEDIA DATABASE MANAGEMENT SYSTEMS

B. Prabhakaran
Department of Computer Science and Engineering
Indian Institute of Technology, Madras, India
and
University of Maryland at College Park, Maryland, USA

Library of Congress Cataloging-in-Publication Data

A C.I.P. Catalogue record for this book is available from the Library of Congress.

Prabhakaran: Multimedia Database Management Systems

First Indian Reprint 2007
Third Indian Reprint 2009

ISBN 978-81-8128-652-9

This edition is manufactured in India for sale in India, Pakistan, Bangladesh, Nepal and Sri Lanka and any other country as authorized by the publisher.

This edition is published by Springer (India) Private Limited, (a part of Springer Science+Business Media, Registered Office: 3rd Floor, Gandharva Mahavidyalaya, 212, Deen Dayal Upadhyaya Marg, New Delhi – 110 002, India.

Printed in India by Rashtriya Printers, Delhi.

CONTENTS

PREFACE

Multimedia databases are very popular because of the wide variety of applications that can be supported. These applications include Video-on-Demand (VoD), teaching aids, multimedia document authoring systems, and shopping guides amongst a score of many others. Multimedia databases involve accessing and manipulating stored information belonging to different media such as text, audio, image, and video. The distinctions between multimedia databases and the traditional ones are due to the following characteristics of media objects :

- Sizes of the media objects (in terms of bytes of information)
- Real-time nature of the information content
- Raw or uninterpreted nature of the media information.

These characteristics in turn raise the following issues:

1. Storage of media objects needs different techniques due to the volume as well as the real-time requirement for their fast retrieval.
2. The contents of media objects are largely binary in nature. Hence, they have to be interpreted based on the type of media, contents of the objects, and the needs of an application. As an example, a facial image will be stored as a binary file. Interpretations have to be made for identifying the features of a face such as color of hair, eyes, shape of nose, etc. These interpretations, termed *metadata*, have to be automatically or semiautomatically generated from media objects.
3. Fast access to stored multimedia information requires different indexing techniques to be provided for handling various media objects.
4. Media objects, associated metadata, the objects' temporal and spatial characteristics have to be modeled in such a way that they can be easily manipulated.

5. Accessing multimedia information is done through user queries that describe the metadata associated with the objects as well as the objects' temporal and spatial characteristics.

6. Multimedia information can be distributed over computer networks. Accessing distributed multimedia data necessitates support from network service provider for communicating large media objects with real-time requirements.

Our aim in this text is to bring out the issues and the techniques used in building multimedia database management systems. The book is organized as follows. In Chapter 1, we provide an overview of multimedia databases and underline the new requirements for these applications. In Chapter 2, we discuss the techniques used for storing and retrieving multimedia objects. In Chapter 3, we present the techniques used for generating metadata for various media objects. In Chapter 4, we examine the mechanisms used for storing the index information needed for accessing different media objects.

In Chapter 5, we analyze the approaches for modeling media objects, their temporal and spatial characteristics. Object-oriented approach, with some additional features, has been widely used to model multimedia information. We discuss two systems that use object-oriented models : *OVID* (Object Video Information Database) and *Jasmine*. Then, we study the models for representing temporal and spatial requirements of media objects. We also describe authoring techniques used for specifying temporal and spatial characteristics of multimedia databases. In Chapter 6, we explain different types of multimedia queries, the methodologies for processing them and the language features for describing them. We also study the features offered by query languages such as SQL/MM (Structured Query Language for Multimedia), PICQUERY$^+$, and Video SQL. In Chapter 7, we deal with the communication requirements for multimedia databases. A client accessing multimedia data over computer networks needs to identify a schedule for retrieving various media objects composing the database. We identify possible ways for generating a retrieval schedule. In Chapter 8, we tie together the techniques discussed in the previous chapters by providing a simple architecture of a distributed multimedia database management system.

The book can be used as a text for graduate students and researchers working in the area of multimedia databases. It can also be used for an advanced course for motivated undergraduates. More over, it can serve as a basic reading material for computer professionals who are in (or moving to) the area of multimedia databases.

Acknowledgment

I would like to thank Prof V.S. Subrahmanian for his encouragement. Thanks to Selcuk for his meticulous reviews and to Eenjun for his feedback. I have benefitted a lot by interacting with them. I learnt a lot by working with Prof S.V. Raghavan and I thank him for that. I acknowledge Prof R. Kalyanakrishnan for his moral support and encouragement. Thanks to Prof. P. Venkat Rangan for his support in many instances.

Thanks to my motivating parents, Balakrishnan and Saraswathi, for their love and constant encouragement. Special thanks to my wonderful wife, Raji, for her love, kindness, patience, and encouragement. That she could pitch in with her reviews of the book was really nice. My son, Gokul, brought an entirely new dimension to our life. His loving company and his playfulness have brought lots of joy and happiness in our life. He even co-authored the book by his bangings on the keyboard. Though I would like to attribute any mistakes in the book to his co-authorship, Raji would not let me do so. I acknowledge the love and support provided by my brothers, Sridhar and Shankar, Manni, the loving kids Madhu and Keechu.

Finally, the research work for writing this book was supported by the Army Research Office under grant DAAH-04-95-10174, by the Air Force Office of Scientific Research under grant F49620-93-1-0065, by ARPA/Rome Labs contract Nr. F30602-93-C-0241 (Order Nr. A716), Army Research Laboratory under Cooperative Agreement DAAL01-96-2-0002 Federated Laboratory ATIRP Consortium and by an NSF Young Investigator award IRI-93-57756.

B. Prabhakaran

1

INTRODUCTION

Multimedia databases can support a variety of interesting applications. Video-on-Demand (VoD), teaching aids, multimedia document authoring systems, and shopping guides are examples of these applications. Multimedia databases deal with storage and retrieval of information comprising of diverse media types such as text, audio, image, and video. The following characteristics of media objects influence multimedia database management systems.

- **Large sizes :** This influences the storage and retrieval requirements of media objects. In the case of distributed multimedia databases, the communication requirements also depend on the sizes of the objects.
- **Real-time nature :** This factor along with the sizes of the objects influence the storage and communication requirements.
- **Raw or uninterpreted nature of information :** Contents of the media objects such as audio, image, and video are binary in nature. Hence, multimedia databases have to derive and store interpretations about the contents of these objects.

In this chapter, we consider typical multimedia database applications and discuss how traditional database management functions such as storing, modeling, accessing, and querying have to be reconsidered for handling multimedia objects.

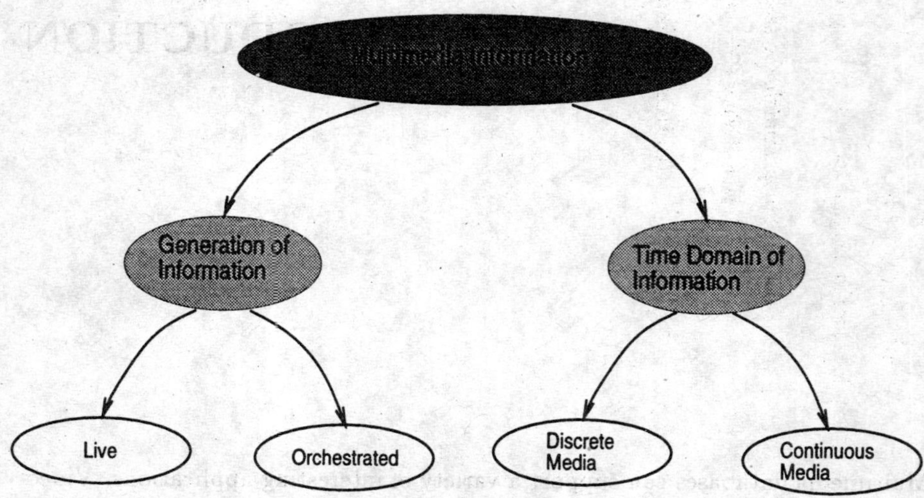

Figure 1.1 Classification of Multimedia Information

1.1 TYPES OF MULTIMEDIA INFORMATION

Multimedia information may be classified depending either on the mode of generation or on the corresponding time domain, as shown in Figure 1.1. The generation of multimedia objects can be either through multimedia devices such as video cameras or through accessing multimedia databases. Based on the generation methodology, multimedia information can be classified as :

- **Orchestrated** : Here, the capture and/or generation of information is done by retrieving stored objects. Stored multimedia lecture presentations, on-demand servers, and other multimedia database applications fall under this category.

- **Live** : Here, information is generated from devices such as video camera, microphone or keyboard. Multimedia teleconferencing and panel discussion applications fall under this category. In these applications, participants communicate among themselves by exchanging multimedia information generated from video camera or microphone.

Multimedia information can be classified into the following categories with respect to the time domain.

- **Discrete (or Time independent) media** : Media such as text, graphics and image, have no real-time demands. Hence, they are termed discrete media.
- **Continuous (or Time dependent) media** : In continuous media, information becomes available at different time intervals. The time intervals can be periodic or aperiodic depending on the nature of the media. Audio and video are examples of periodic, continuous media.

Orchestrated and live multimedia applications can be composed of both discrete and continuous media. In a live multimedia presentation, images generated using document cameras fall under the discrete media category whereas information generated from video camera and microphone fall under the continuous media category. In live applications, temporal relationships of the objects in a media are implied. These temporal relationships are related to the sampling rate used for the media. For video, it is 30 frames/second in the United States and 25 frames/second in the Europe. For audio, the rate at which information is acquired varies from 16 Kbps to 1.4 Mbps.

In a similar manner, orchestrated applications are composed of both discrete and continuous media. The difference is that in the case of an orchestrated multimedia application, temporal relationships for various media objects have to be explicitly formulated and stored. These temporal relationships describe the following :

- When an object should be presented
- How long it should be presented
- How an object presentation is related to those of others (for example, audio object might have to be presented along with the corresponding video).

1.2 MULTIMEDIA DATABASE APPLICATIONS

Multimedia databases are orchestrated applications where objects are stored and manipulated. Many application scenarios involve storage, processing, and retrieval of multimedia data. We can consider the following applications.

Video-on-Demand (VoD) Servers : These servers store digitized entertainment movies and documentaries, and provide services similar to those of a videotape rental store. Digitized movies need large storage spaces and hence these servers typically use a number of extremely high capacity storage devices, such as optical disks. Users can access a VoD server by searching on stored information such as video's subject title, and can have a real-time playback of the movie.

Multimedia Document Management Systems : This is a very general application domain for multimedia databases. It involves storage and retrieval of multimedia objects which are structured into a multimedia document. The structuring of objects into a multimedia document involves specifying the following :

- The temporal relationships among the objects composing the multimedia document
- The spatial relationships that describe how objects are to be presented on a monitor

Multimedia document management systems can have applications in technical documentation of product maintenance, education, and geographical information systems. These applications use objects such as images, video and audio to a large extent. These objects along with some useful text can be structured into a multimedia document. An interesting aspect of multimedia documents is that media objects can be distributed over computer networks. Authors can work in a collaborative manner to structure the data into a multimedia document.

Multimedia Mail Systems : They integrate features, such as multimedia editing and voice mail, into the traditional electronic mailing systems. The messages, composed of multimedia objects, are forwarded to the recipients.

Multimedia Shopping Guide : It maintains huge amounts of shopping information in the form of a multimedia document. The information may be about products, stores, ordering, etc. Customers can dial up a retail store, look at products of interest, and order them over computer networks (and pay for the products, if the network offers secure services).

1.2.1 Multimedia Database Access : An Example

Consider a video-on-demand (VoD) database management system with a repository of large number of movies. Customers can access the VoD server, download and watch movies. A client can query the server regarding the available movies. The VoD server can store the following information about the available movies :

- A short video clip of the movie
- An audio clip associated with the video clip
- Two important still images taken from the movie
- Text, giving the details such as the director, actors, actresses and other special features of the movie

A client can query the VoD database in many possible ways. For instance, consider the following customer queries :

Query 1 : What are the available movies with computerized animation cartoons?

VoD Server Response : The VoD server shows the details regarding the movies : *Who Framed Roger Rabbit* and *Toy Story*.

Query 2 : Show the details of the movie where a cartoon character speaks *this* sentence. (*This* sentence is an audio clip saying : '*Somebody poisoned the water hole*').

VoD Server Response : The server shows the clip from the movie *Toy Story* where the cartoon character Woody speaks the above sentence. The response comprises of video and audio clips, associated still images and text.

Query 3 : Show the movie clip where the following video clip occurs : the cartoon character *Wooody* sends its Green Army men on a *recon mission* to monitor the gifts situation on its owner's birthday.

VoD Server Response : The server shows the requested clip from the movie *Toy Story* along with associated audio, still images and text.

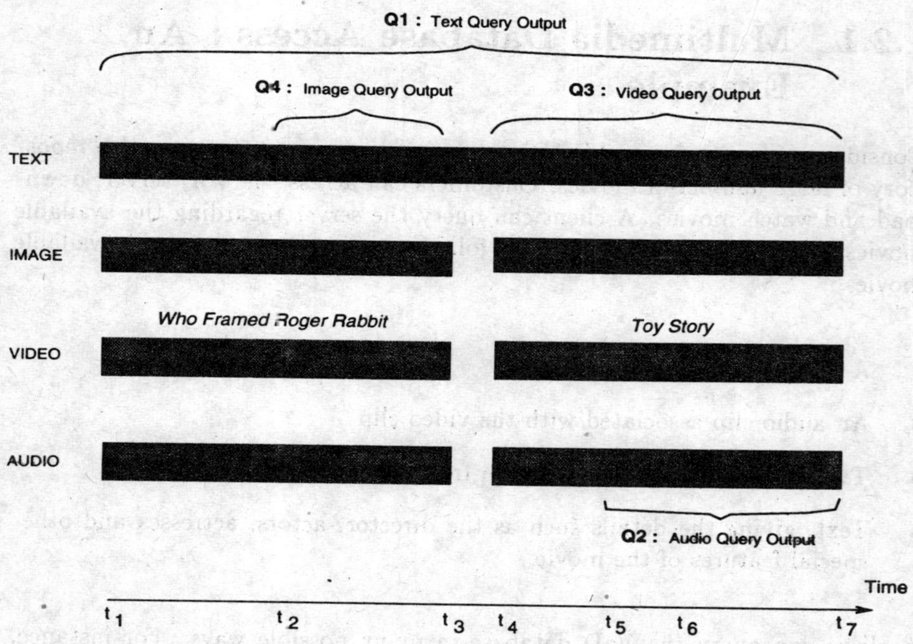

Figure 1.2 VoD Server : Example Queries and Output

Query 4 : Show the details of the movie where *this* still image appears as part of the movie. (*This* image describes the scene where the cartoon character *Jessica Rabbit* is thrown from the animated cab).

VoD Server Response : The server shows the still image from the movie *Who Framed Roger Rabbit* as well as the associated details of the movie.

The customer can give a combination of the above queries also. Depending upon the nature of the query, the multimedia objects composing the response varies. Figure 1.2 shows the objects to be presented for the queries discussed above. For instance, the response to query 1 is composed of objects $W, X1, X2, X3, X4, Y1, Y2, Z1, Z2$ whereas the response for query 2 is composed of objects $X3, X4, Y2$ and portions of objects $W \& Z2$.

1.3 MULTIMEDIA OBJECTS : CHARACTERISTICS

As can be seen from the above example, multimedia databases can be accessed by queries on any of the objects composing the databases. The properties of these media objects distinguish the needs of a multimedia database management system from those of a traditional one, as discussed below.

Text Data : is often represented as strings. However, text, as used in multimedia document systems, includes structural information such as title, author(s), authors' affiliations, abstract, sections, subsections, and paragraphs. Hence, one needs a language environment to reflect the structural composition of the text data. *Standard Generalized Markup Language* (SGML) is a document representation language defined by the International Standards Organization (ISO). Another environment, named the *Hypermedia/Time-based Structuring Language* (HyTime), has also been defined to include support for hypermedia documents (hypertext with multimedia objects), with links and support for inclusion of multimedia objects in a text document specification. SGML together with HyTime can be used for developing multimedia documents.

Audio Data : has an inherent time dependency associated with it. The time scale associated with audio objects has to be uniform for a meaningful interpretation. Audio has to be digitized before it can be processed. Size of digitized audio depends on the technique used, which in turn depends on the desired audio quality. For example, a *normal* voice quality digitization is done at 8 KHz with 8 bits per sample, and hence it produces 64 Kb/s of data. CD quality digitization is carried out at 44.1 KHz sampling rate with 16 bits per sample and hence produces 1.4 Mb/s. Digitized audio can be effectively compressed to reduce storage requirements.

Image Data : represents digitized drawings, paintings, or photographs. Size of a digitized image depends on the required quality. Color images and photographs require more storage space. Typically, a color image or a photograph needs the RGB (Red, Green and Blue) components of each pixel to be stored. Depending on the color scale chosen, one might need 8 bits per color component implying 24 bits per pixel. Hence, for a 1024 * 1024 pixel image, a storage space of 24 Mbits is needed. Compression schemes are used to reduce the volume of data that needs to be stored. Most compression schemes employ algorithms that exploits the redundancy in the image content. Different compression algorithms as well as storage representations can be employed and this results in different *formats* of the digitized images and photographs. *Joint Photographers Encoding Group* (JPEG) is one such format for images, which has been standardized by the ISO. Other popular formats include *Graphic Interchange Format* (GIF) and *Tag Image Format* (TIFF).

Graphics Data : represents the concepts that allow generation of drawings and other images based on formal descriptions, programs, or data structures. International standards have been specified for graphics systems to serve as a basis for industrial and scientific applications.

Video Data : represents the time dependent sequencing of digitized pictures or images, called *video frames.* The number of video frames for one second depends on the standard that is employed. NTSC (National Television System Committee) standard employs 30 frames/second while PAL (Phase Alternation Line) standard employs 25 frames/second. Also, the pixel size of a frame depends on the desired quality. Normal NTSC frames are 512 * 480 pixels in size. HDTV (High Definition Television) frames employ 1024 * 1024 pixels. The number of bits needed per pixel reflects the quality of digitized video frame. Digitized video requires large storage space. Compression schemes need to be employed to reduce the volume of data to be stored. *Motion Pictures Encoding Group* (MPEG) is a standard that has been specified by the ISO for compression and storage of video. The standard *MPEG 2* specifies the methodology for storing audio along with compressed video.

Generated Media : represents computer generated presentations such as animation and music. Generated media differs from others in the sense that data is generated based on a standard representation. As an example, Musical Instrument Digital Interface (MIDI) defines the format for storing and generating music in computers.

1.3.1 Access Dimensions of the Media Objects

With reference to the process of accessing the contents, media objects can be considered as one of the following :

1-dimensional Objects : Text and audio have to be accessed in a contiguous manner (as ASCII strings or signal waves), as shown in Figure 1.3 (a). Hence, text and speech can be considered as *1-dimensional* objects.

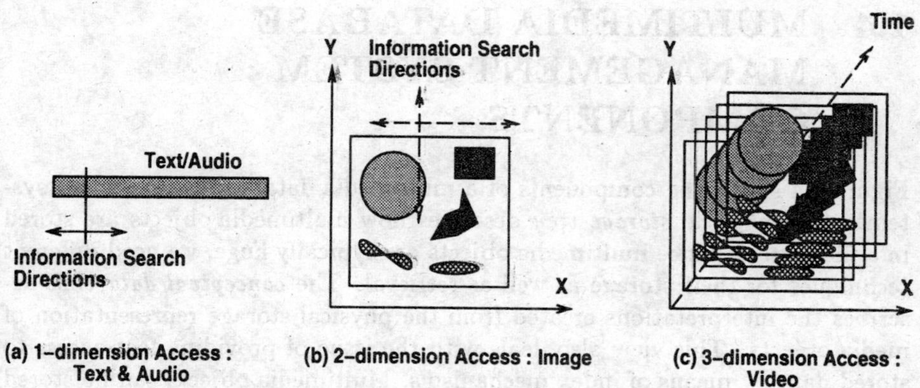

Figure 1.3 Access Dimension of Media Objects

2-dimensional Objects : Access to image data can be done with reference to the spatial locations of objects. For example, a query can search for an object that is to the *right of* or *below* a specified object. So, image objects can be considered as *2-dimensional*, since it has a spatial content as shown in Figure 1.3 (b).

3-dimensional Objects : Video has spatial characteristics as well as temporal characteristics as shown in Figure 1.3 (c). Access to video can be done by describing the temporal as well as the spatial content. For example, a query can ask for a movie to be shown from *10 minutes after its starting point.* Hence, video can be considered as a *3-dimensional* object.

The access dimension of an object, in a way, describes the complexity in the process of searching. For 1-dimensional objects, such as text and audio, the access is limited to the keywords (or other related details) that appears as part of text or speech. For images, the access is done by specifying the contents as well as their spatial organization. In a similar manner, access to video should comprise of the sequencing of video frames in the time domain. In the following sections, we discuss how the characteristics of media objects influence the components of a multimedia database management system.

1.4 MULTIMEDIA DATABASE MANAGEMENT SYSTEM : COMPONENTS

Figure 1.4 shows the components of a multimedia database management system. The *physical storage view* describes how multimedia objects are stored in a file system. Since multimedia objects are typically huge, we need different techniques for their storage as well as retrieval. The *conceptual data view* describes the interpretations created from the physical storage representation of media objects. This view also deals with the issue of providing fast access to stored data by means of index mechanisms. Multimedia objects can be stored in different systems and users might access stored data over computer networks. This leads to a *distributed view* of multimedia databases.

Users can query multimedia databases in different ways, depending on the type of information they need. These queries provide a *filtered view* of the multimedia databases to the users by retrieving only the required objects. The objects retrieved from the database(s) have to be appropriately presented, providing the *user's view* of the multimedia database. Though these views are true for a traditional database management system, the diverse characteristics of media objects introduce many interesting issues in the design of a multimedia database management system, as discussed below.

1.4.1 Physical Storage View

The main issue in the physical storage of multimedia objects is their sizes. Sizes of objects influences both the storage capacity requirements and the retrieval bandwidth (in terms of bits per second) requirements. Table 1.1 describes the size and the retrieval disk bandwidth requirements for different media, based on their format of representation. The disk bandwidth requirements of discrete media such as text and images depend on a multimedia database application. This is because these media do not have any inherent temporal requirements. The bandwidth requirements of discrete media might depend on the number of images or the number of pages of text, that an application needs to present within a specified interval of time.

On the contrary, continuous media such as video and audio have inherent temporal requirements, e.g., 30 frames/second for NTSC video. These temporal requirements imply that an uncompressed 5 minutes video clip object will re-

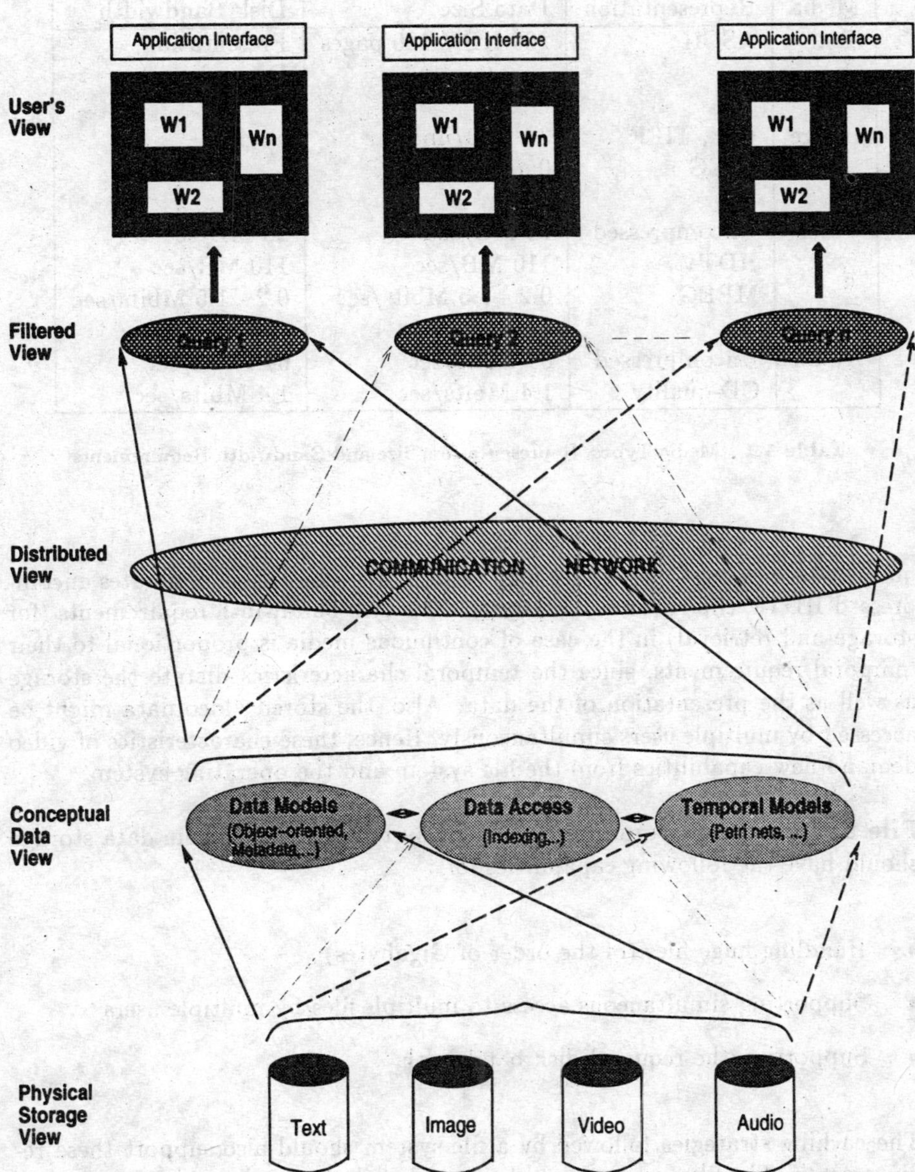

Figure 1.4 Components Involved in Multimedia Databases

Media	Representation	Data Size	Disk Bandwidth
Text	ASCII	200 KB / 100 pages	Presentation Dependent
Image	GIF, TIFF JPEG	3.2 MB/image 0.4 MB/image	— do —
Video	Uncompressed HDTV MPEG	20 MB/sec 110 MB/sec 0.2 - 1.5 Mbits/sec	20 MB/sec 110 MB/sec 0.2 - 1.5 Mbits/sec
Audio	Uncompressed CD-quality	64 Kbits/sec 1.4 Mbits/sec	64 Kbits/sec 1.4 Mbits/sec

Table 1.1 Media Types, Representation, Size and Bandwidth Requirements

quire 300 times its storage space for 1 second. For example, a 5 minutes uncompressed HDTV clip requires 33 GBytes. The disk bandwidth requirements (for storage and retrieval) in the case of continuous media is proportional to their temporal requirements, since the temporal characteristics dictate the storage as well as the presentation of the data. Also, the stored video data might be accessed by multiple users simultaneously. Hence, these characteristics of video demand new capabilities from the file system and the operating system.

File System Requirements : A file system for multimedia data storage should have the following capabilities for :

- Handling huge files (of the order of Gigabytes)
- Supporting simultaneous access to multiple files by multiple users
- Supporting the required disk bandwidth

The caching strategies followed by a file system should also support these requirements. The file system might have to distribute the data over an array of disks in the local system or even over a computer network. Also, the file system can provide new application programming interfaces apart from the traditional ones such as *open, read, write, close* and *delete.* The new application programming interfaces can support *play, fast forward* and *reverse* for continuous media such as video.

Operating System Requirements : Operating system supporting multimedia applications should have capabilities for handling real-time characteristics. This necessitates that an operating system addresses the following issues.

- Scheduling of application processes
- Communication between an application process and the operating system kernel

The scheduling policy followed by the operating system should allow for the real-time characteristics of multimedia applications. For real-time scheduling, the operating system might have to reserve the resources required for an application process. This implies that depending on the availability of resources, an application process may or may not be admitted for execution by the operating system. Also, a general purpose operating system will have a mixture of processes running with and without real-time requirements. Hence, there is a need for more than just one scheduling policy. Another important required feature is the reduced overhead in the communication between application processes and the operating system kernel. This overhead directly affects the performance of applications.

1.4.2 Conceptual Data View

Physical storage of multimedia objects deals with raw digitized data. In this stage, multimedia objects are in binary form. These objects are acquired (from devices) and created (digitized, compressed and stored) independent of its contents. For using these objects as meaningful data, one needs to identify their *content.* The description of the objects' content, called *metadata*, is subjective in nature and is dependent on the media type as well as the role of an application. As an example, consider the facial image of a person. The description of the person's nose (long, short, slanted, sharp, etc.,) is subjective. The description also depends on the role of the application. Feature description of a facial image may not be needed for a particular application and hence the database may not carry such descriptions. In a similar manner, the metadata associated with a video clip is subjective and depends on the role of an application. Meaningful descriptions of video clips have to be identified and stored in the database.

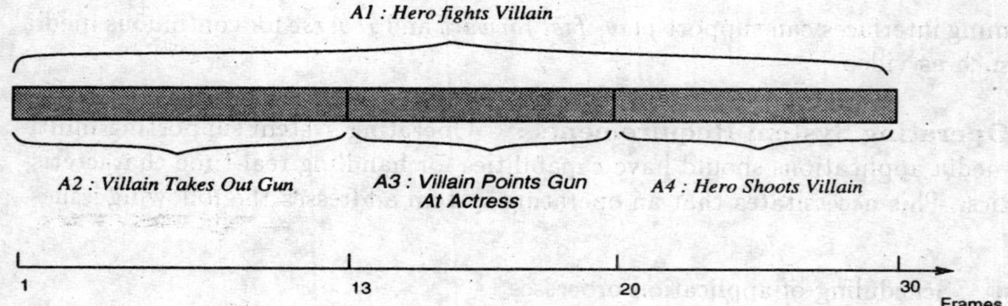

Figure 1.5 Example Description of a Video Clip

As an example, consider a video clip of a movie. The sequence of frames contains actors, actresses, the background of the scene, action going on in the scene, etc. Hence, a description of the video clip might contain descriptions of the characters acting in the movie, the background and the action part of it. As shown in Figure 1.5, the action part of it might be described based on the theme of the video clip : *hero fights the villain* (action *A1*). The metadata can also be more descriptive like : *the villain takes out a gun from his pocket* (action *A2*), *villain points the gun at the actress* (action *A3*) and *the hero shoots the villain* (action *A4*).

The conceptual data view of raw multimedia data helps in building a set of abstraction. These abstractions form a data model for a particular application domain. For fast accesses, we need indexing mechanisms to *sort* the data according to the features that are modeled. A multimedia database may be composed of multiple media objects whose presentation to the user has to be properly synchronized. This synchronization characteristics are described by temporal models. Hence, the conceptual view of multimedia data consists of the following components :

- Metadata
- Indexing mechanisms
- Temporal models
- Spatial models
- Data models

Metadata : deals with the content, structures, and semantics of media objects. The creation of metadata depends on the media type and the type of information which an application wants to describe as part of the metadata. From the maintenance of multimedia database point of view, it is important that techniques for automatic (or semi-automatic) generation of metadata for each media type are available. For video media, the techniques should identify camera shots, characters in a shot, background of a shot, etc. Human interaction might be needed to annotate the sequences based on their semantic content, thereby rendering the techniques semi-automatic. For image data, techniques should extract and describe the features of interest. In a similar manner, recognition techniques might be needed for identifying keywords in audio and text data.

Indexing Mechanisms : Multimedia databases need indexing mechanisms to provide fast access. The techniques developed for traditional databases do not serve this purpose fully, since new object types have to be dealt with. The indexing mechanisms should be able to handle different features of objects such as color or texture.

Temporal Models : describe the time and duration of presentation of each media object as well as their temporal relationships to other media objects. For instance, Figure 1.2 describes the temporal relationship among the objects composing a VoD database. Here, as an example, the video object $Y1$ has to be presented at time $t1$ for a duration of $t3 - t1$ and has to be synchronized with the presentation of audio object $Z1$. The temporal requirements of objects composing a multimedia database have to be specified appropriately.

Spatial Models : represent the way media objects are presented, by specifying the layout of windows on a monitor. Figure 1.6 shows a possible organization of windows for presenting the objects in the VoD database discussed in Section 1.2.1.

Data Models : Object-oriented approach is normally used to represent the characteristics of objects, metadata associated with them, their temporal and spatial requirements.

The influence of the media characteristics on the conceptual data view of a multimedia database management system is summarized in Table 1.2.

Media Characteristics	Conceptual Data View Requirements
Raw, un-interpreted data	Creation of metadata and data models
Fast access to media information	Indexing mechanisms
Multiple media objects in database	Temporal models to represent synchronization of presentation, Spatial representation of windows

Table 1.2 Media Characteristics and Conceptual Data View Requirements

1.4.3 Distributed View

Like any other information, multimedia data can also be distributed over computer networks. Huge sizes of media objects require large bandwidths or *throughput* (in terms of bits per second). Real-time nature of the objects needs guarantees on end-to-end delay and delay jitter. End-to-end delay specifies the maximum delay that can be suffered by data during communication. Delay jitter describes the *variations* in the end-to-end delay suffered by the data. Guarantees on end-to-end delay and delay jitter are required for *smooth* presentation of continuous media objects such as audio and video. For example, if video data is not delivered in periodic intervals (within the bounds specified by the delay jitter parameter), users may see an unpleasant, jerky video presentation.

Applications such as collaborative multimedia document authoring applications might involve simultaneous communication among different entities (*viz.*, application processes and computer systems). Hence, they might need a group of channels for communication. Existing communication protocols address the needs of more traditional applications such as file transfer, remote login, and electronic mail. These applications do not have any real-time requirements and so there is little need for large bandwidths (though the amount of information to be transferred can be huge). So, distributed multimedia applications require a new generation of protocols.

Client retrieving information from a multimedia database server needs to identify when the objects are needed for their presentation. The times of objects'

Media Characteristics	Demands on the Network Provider
Huge Data Size	Large Communication Bandwidth
Real-time Nature	Guaranteed Bandwidth, Delay, and Delay Jitter
Data and User Distribution on the Network	Grouped channels, Retrieval Schedule

Table 1.3 **Media Characteristics and Communication Requirements**

presentations are described by their temporal relationships. Due to their huge sizes, not many media objects can be buffered by the client. Also, the bandwidth offered by the network is not unlimited. Hence, based on the temporal relationships, the buffers required and the available network bandwidth, the client needs to identify a *retrieval schedule* for requesting objects from the server. As an example, consider the temporal relationship shown in Figure 1.2 for a VoD server application. Assuming that the objects are distributed, we need to identify a retrieval schedule for image objects (and similarly, for other media objects) so that $X1$ can be presented at t_1, $X2$ at t_2, $X3$ at t_4, and $X4$ at t_5.

Table 1.3 summarizes the communication requirements of typical multimedia database applications.

1.4.4 Filtered View

Filtered view of a multimedia database is provided by a user's query to get the required information. The query can be on any of the media that compose a database, as discussed in Section 1.2.1. User's query can be of the following types :

- Query on the content of media objects
- Query by example (QBE)
- Time indexed queries

- Spatial queries
- Application specific queries

Content Based Queries : Queries on the content of media objects typically require a search on the metadata associated with the objects. Queries 1 and 3 discussed in Section 1.2.1 belong to this category.

Query By Example : Considering the VoD server application, users can make queries by examples such as :

- Get me the movie in which *this* scene (an image) appears
- Get me the movie where *this* video clip occurs
- Show me the movie which contains *this* song

In these examples, the italicized *this* refers to the multimedia object that is used as an example. The multimedia database management system has to process the *example* data (*this* object) and find one that matches it, i.e., the input query is an object itself. The requirement for similarity can be on different characteristics associated with the media object. As an example, for image media, similarity matching can be requested on texture, color, spatial locations of objects in the example image, or shapes of the objects in the example image. The required similarity matching between the queried object and database objects can be exact or partial. In the case of partial matching, we need to know the *degree of mismatch* that can be allowed between the example object and the ones in the database.

Time Indexed Queries : Since multimedia databases are composed of time-dependent or continuous media, users can give queries in the temporal dimension as well. For example, a time indexed query for a VoD server can be: Show me the movie *30 minutes after* its start.

Spatial Queries : Media objects such as image and video have spatial characteristics associated with them. Hence, users can issue a query like the following one : Show me the image where President Yelstin is seen to *the left of* President Clinton.

Application Specific Queries : Multimedia databases are highly application specific. Queries, therefore, can be application specific too. We can

Media Characteristics	Filtered view requirements
Binary nature of data	Content based queries, Query by examples
Temporal nature of data	Time indexed queries
Spatial nature of data	Spatial queries
Diverse application requirements	Application specific queries

Table 1.4 Media Characteristics and Querying Requirements

consider databases such as medical or geographic information database. Users can ask queries such as :

- Show me the video where the tissue *evolves* into a cancerous one
- Show me the video where the river *changes its course*

As discussed above, user queries can be of different types. Hence, the query processing strategies and the query language features have to address the specific requirements of the corresponding multimedia database applications. Table 1.4 summarizes the requirements on the filtered view of a multimedia database management system.

1.4.5 User's View

User's view of a multimedia database management system is characterized by the following requirements.

- User query interface
- Presentation of multimedia data
- User interaction during presentation

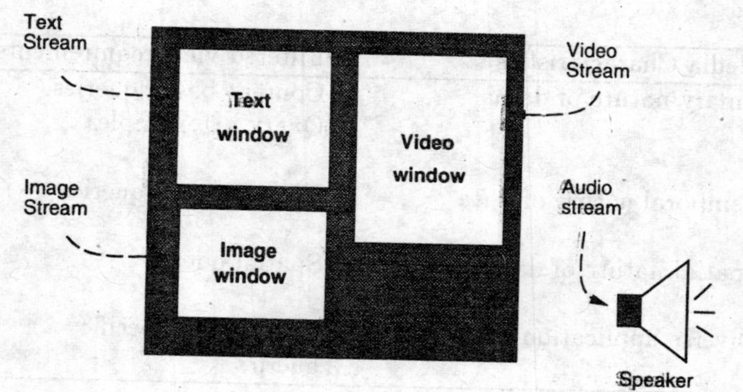

Figure 1.6 Presentation of the Multimedia Information to User

User Query Interface : Query interface should allow users to query by content, example, time, space, or a combination of these possibilities. For queries by example, the user query interface has to obtain the example object from appropriate devices (e.g., example image object can be obtained through a scanner or from a stored file). The query interface can provide suggestive inputs so as to ease the process of querying. Also, in case of partial matching of the resolved queries, the query interface can suggest ways to modify the query to get exact matches.

Presentation of Multimedia Data : Media objects can be of different formats. For example, images can be stored in *tiff* or *gif* format. Hence, the object presentation tools should be capable of handling different formats. In some cases, there might be a necessity to convert data from one format to another before presentation. Also, multimedia objects composing the responses to user's queries have associated temporal constraints. These constraints specify the time instants and the durations of the presentations of various multimedia objects. In the example discussed in Figure 1.2, the temporal constraint is indicated by the time axis along with the respective time marks. In a similar manner, the presentation of multimedia objects may have spatial constraints. These constraints describe the layout of windows on the user's screen for presentation of different objects. Figure 1.6 shows a possible spatial organization for presenting the retrieved multimedia database information in the VoD server example discussed in Section 1.2.1.

User Interaction During Presentation : User can interact during the presentation of multimedia objects. The interaction is more complex (compared to that in the traditional databases) since multiple media objects are involved.

Media Characteristics	User's View Requirements
Different media representation formats	Different presentation tools
Different types of queries	Different query interfaces
Simultaneous presentation of multiple media objects	Handling user interaction on the objects' presentation

Table 1.5 Media Characteristics and User's View Requirements

For example, devices such as microphone and video camera can be used for speech and gesture recognition, apart from the traditional ways of handling inputs from keyboard and mouse. Hence, simultaneous control of different devices and handling of user inputs is required. The input from the user can be of following types :

- Modify the quality of the presentation, (e.g) reduction or magnification of the image
- Direct the presentation, (e.g.) skip, reverse, freeze or restart

The requirements on the user's view as influenced by the media characteristics are summarized in Table 1.5.

1.5 CONCLUDING REMARKS

Multimedia database management system is an orchestrated application where stored objects are accessed and manipulated through queries. This stored and query based access model gets complex because of the diverse characteristics of various media objects. The media characteristics that influence the requirements of multimedia database management systems are :

- Sizes of the objects,
- Real-time nature,

- Raw or un-interpreted nature of the information.

These media characteristics influence the following components of the multimedia database management system :

- Storage of media objects : the *physical storage view.*
- Interpretation of the raw information : the *conceptual data view.*
- Physical location of media objects and users : the *distributed view.*
- Querying databases : the *filtered view.*
- Interfacing multimedia database applications to users : the *user's view*

Table 1.6 summarizes the media characteristics and the requirements of a typical multimedia database management system.

Bibliographic Notes

An overview of multimedia systems can be found in [114, 107]. Issues in providing live multimedia applications such as multimedia conferencing are discussed in [66, 62, 64, 68, 81, 100].

The features of Standard Generalized Markup Language (SGML) are described in [23, 70, 143]. The *Hypermedia/Time-based Structuring Language* (HyTime), has been defined to include support for hypermedia documents (hypertext with multimedia objects) and the details can be found in [72]. Discussions on hypermedia and world-wide web appear in [166, 167, 168, 147, 67].

Joint Photographers Encoding Group (JPEG) standard has been discussed in [71, 49]. International standards have been specified in [17] for graphics systems to serve as a basis for industrial and scientific applications. *Motion Pictures Encoding Group* (MPEG) standard for compression and storage of video car be found in [105].

MMDBMS View	Media Characteristics	New Requirements
Physical View	Sizes of objects	Management of huge files
	Real-time characteristics	Huge disk bandwidth, Real-time Scheduling
Data Model View	Binary representation of media objects	Metadata creation, New indexing techniques
	Composition of multiple objects	Temporal, Spatial specifications
Distributed View	Sizes of objects, real-time nature	High throughput, guaranteed delay
	Data and user distribution	Grouped channels, Retrieval Schedule
Filtered View	Binary nature	Content based queries, Query by examples
	Temporal nature	Time indexed queries
	Spatial nature	Spatial queries
	Diverse application requirements	Application specific queries
User View	Data representation formats	Different types of presentation tools
	Different types of queries	Query interface design
	Information Presentation	User interaction like skip, fast forward, etc

Table 1.6 Requirements of A Multimedia Database Management System

2

MULTIMEDIA STORAGE AND RETRIEVAL

Large sizes as well as real-time requirements of multimedia objects influence their storage and retrieval. Figure 2.1 shows a logical scenario of a multimedia server that receives a query from a client, retrieves the required data and passes it back to the client as a possible response. The multimedia server needs to store and retrieve data in such a manner that the following factors are taken care of :

- Rate of the retrieved data should match the required data rate for media objects.
- Simultaneous access to multiple media objects should be possible. This might require synchronization among retrieval of media objects (e.g., audio and video of a movie).
- Support for new file system functions such as fast forward and rewind. These functions are required since users viewing multimedia objects such as video can initiate VCR like functions.
- Multiple access to media objects by different users has to be supported.
- Guarantees for the required data rate must be provided.

2.1 MULTIMEDIA OBJECT STORAGE

Multimedia objects are divided into blocks while storing them on disk(s). Each data block can occupy several physical disk blocks. The techniques used

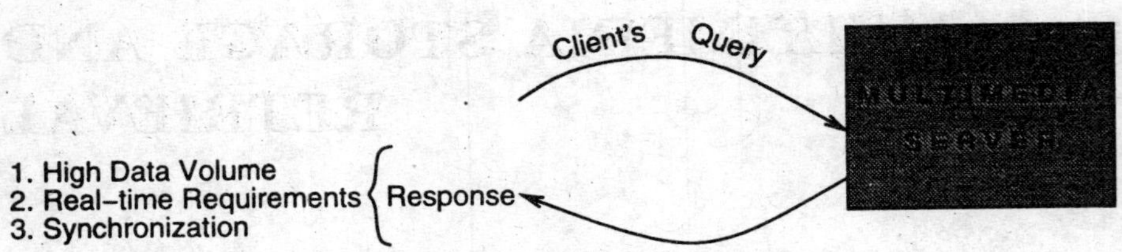

Figure 2.1 Multimedia Data Retrieval

for placing object blocks on a disk influence the data retrieval. The following possible configurations can be used to store objects in a multimedia server.

- **Single Disk Storage :** One possibility is to store objects belonging to different media types in the same disk, as shown in Figure 2.2 (a). If a client's query involves retrieval of multiple objects (belonging to different media), then the multimedia server has to ensure that objects can be retrieved at the cumulative data rate.
- **Multiple Disk Storage :** If multiple disks are available, objects can be distributed across different disks. Figure 2.2 (b) shows one possible approach where individual media objects are stored on independent disks. Since multiple disks are involved, the required rate of data retrieval can be more easily satisfied.
- **Multiple Disks With Striping :** Another possibility while using multiple disks is to distribute the placement of a media object on different disks, as shown in Figure 2.2 (c). The retrieval rate for a media object is greatly enhanced because data for the same object is simultaneously retrieved from multiple disks. This approach, termed *disk striping*, is particularly useful for high bandwidth media objects such as video.

2.1.1 Object Storage On A Single Disk

A media object can be stored entirely on a single disk (as in the case of a single disk server or media-on-a-disk server). Here, the objects have to be stored in such a way that the required retrieval rate is less than the available disk bandwidth. The data blocks can be placed on the disk in the following manner :

- Contiguously

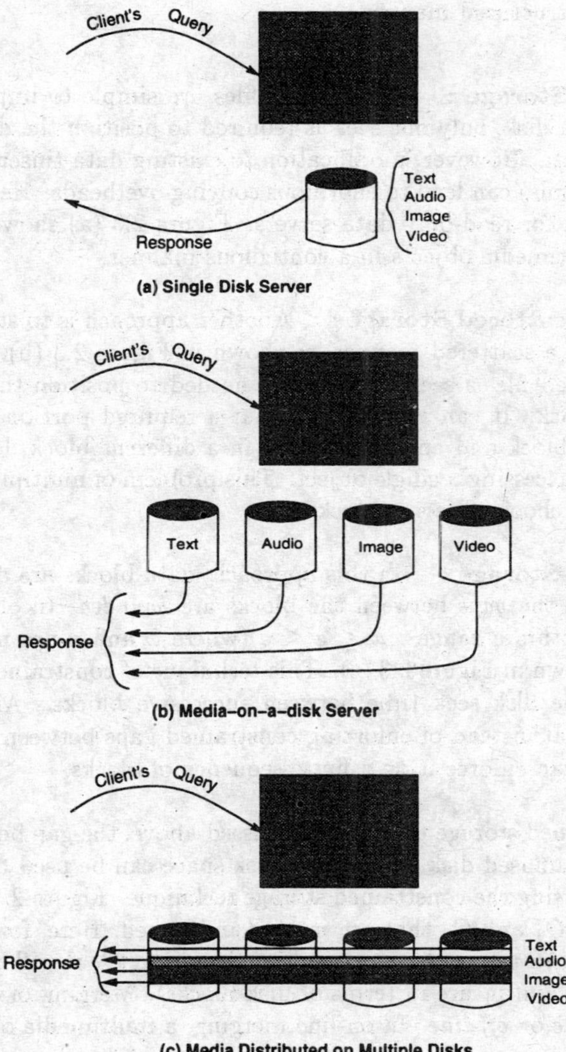

Figure 2.2 Possible Multimedia Data Configurations

- Scattered in a random manner
- Distributed in a constrained manner
- In a log-structured manner

Contiguous Storage : Contiguous files are simple to implement. When reading from a disk, only one seek is required to position the disk head at the start of the data. However, modification to existing data (inserting a chunk of data, for example) can lead to enormous copying overheads. Hence, contiguous files are useful for read-only data servers. Figure 2.3 (a) shows the technique of storing multimedia objects in a contiguous manner.

Randomly Scattered Storage : Another approach is to store multimedia data blocks in a scattered manner, as shown in Figure 2.3 (b). When reading from a scattered file, a seek operation is needed to position the disk head for *every* data block. It can also happen that a required portion of an object is stored in one block and another portion in a different block, leading multiple disk seeks for accessing a single object. This problem of multiple disk seeks can be avoided by choosing larger block sizes.

Constrained Storage : In this approach, data blocks are distributed on a disk such that the *gaps* between the blocks are *bounded.* In other words, gap g has to be within a range : $x \leq g \leq y$ (where x and y are in terms of disk blocks), as shown in Figure 2.3 (c). This technique of constrained storage helps in reducing the disk seek time between successive blocks. Another possible approach is that instead of enforcing constrained gaps between successive pair of blocks, we can enforce it on a finite sequence of blocks.

In the constrained storage technique discussed above, the gap between two data blocks implies unused disk space. This disk space can be used to store another media object using the constrained storage technique. Figure 2.3 (d) shows two media objects O_1 and O_2 that are *merged* and stored. Here, for object O_1, the gap g will be such that $x \leq g \leq y$, and for the object O_2 it will be $x_1 \leq g_1 \leq y_1$ (where x, y, x_1 and y_1 are in terms of disk blocks). Merging of data can either be done *on-line* or *off-line.* In on-line merging, a multimedia object has to be stored with already existing objects. Whereas, in off-line merging, the storage patterns of multimedia objects are adjusted prior to merging.

Log-structured Storage : In log-structured storage, modifications to existing data are carried out in an append-only mode of operation. Figure 2.3 (e) describes the log-structured storage strategy. Here, the modified blocks are

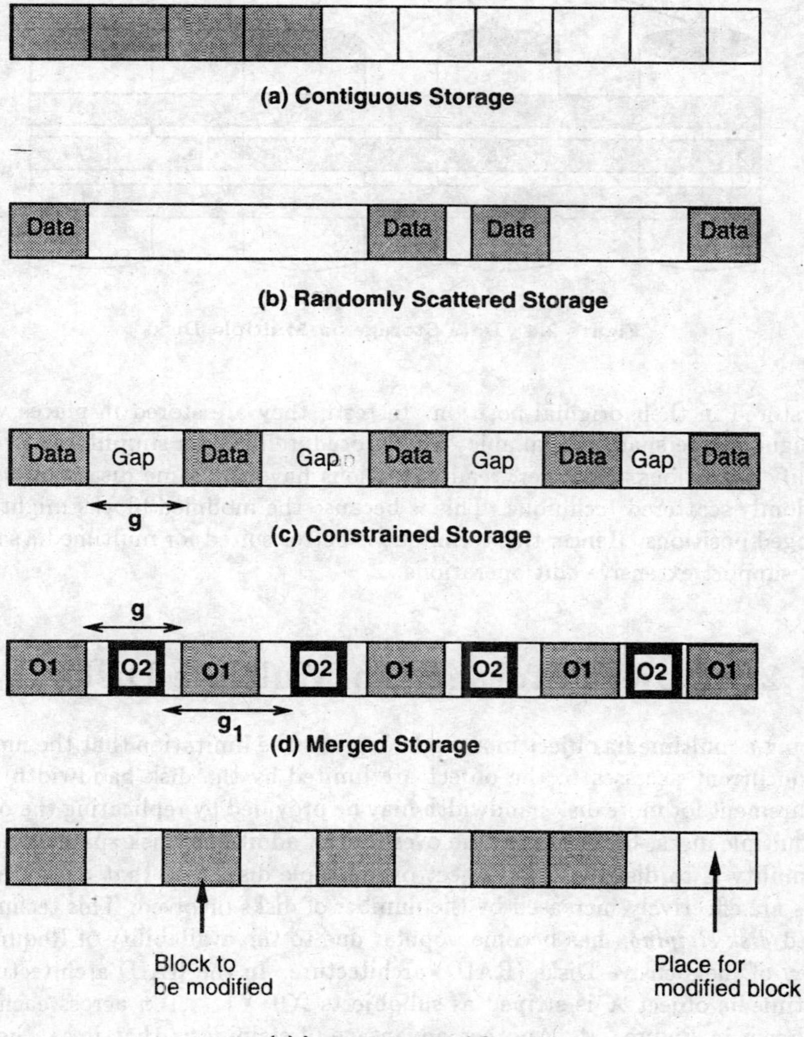

Figure 2.3 Data Storage On A Single Disk

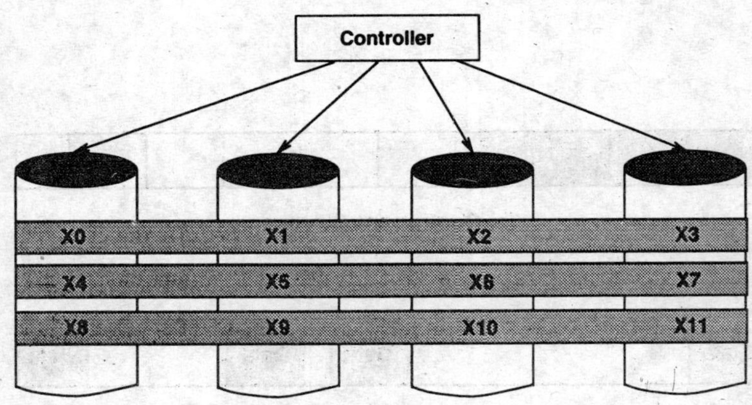

Figure 2.4 Data Storage on Multiple Disks

not stored in their original position. Instead, they are stored in places where contiguous free space is available. This procedure helps in simplifying write or modify operations. However, read operations have the same disadvantages as randomly scattered technique. This is because the modified blocks might have changed positions. Hence, this technique is better suited for multimedia servers that support extensive edit operations.

2.1.2 Object Storage On Multiple Disks

Storing a multimedia object in a single disk has the limitation that the number of concurrent accesses to the object are limited by the disk bandwidth. The requirement for more disk bandwidth may be provided by replicating the object in multiple disks, but it incurs the overhead of additional disk space. Another possibility is to distribute an object on multiple disks, so that data transfer rates are effectively increased by the number of disks involved. This technique, called *disk striping*, has become popular due to the availability of Redundant Array of Inexpensive Disks (RAID) architecture. In the RAID architecture, a multimedia object X is striped as subobjects $X0, X1, ..., Xn$ across each disk as shown in Figure 2.4. Another advantage of striping is that it can help in providing VCR like functions such as fast forward and rewind. For supporting these functions, the retrieval operation can skip a fixed number of subobjects before retrieving the next subobject. For instance, retrieval operation for fast forward can get subobjects $X0, X4$, and $X8$, instead of the whole set of subob-

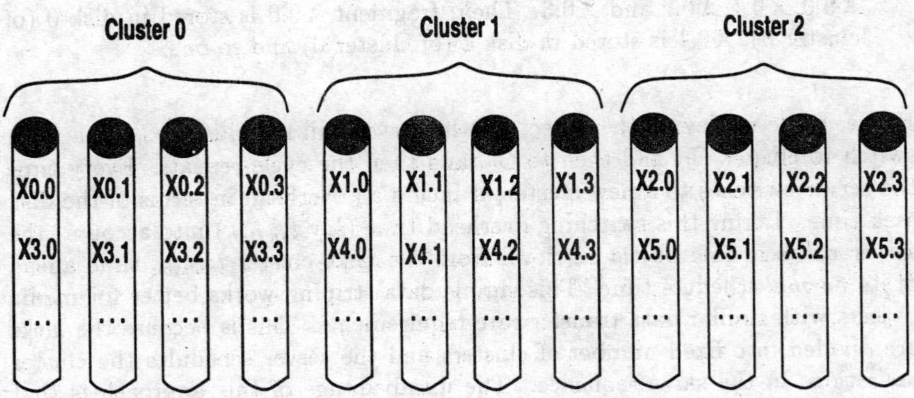

Figure 2.5 Simple Object Striping on Disk Clusters

jects $X0$ through $X11$. Different techniques are used for striping multimedia objects on disks. Here, we discuss the following techniques :

- Simple Striping
- Staggered Striping
- Network Striping

Simple Data Striping : When more number of disks are involved, the disks can be divided into a number of clusters and the data striping can be implemented over the disk clusters, as shown in Figure 2.5. Here, an object is striped as follows :

- First, an object is divided into subobjects. The subobjects are striped *across* disk clusters so that consecutive subobjects of an object X (say, Xi and $Xi+1$) are stored in consecutive clusters and hence in non-overlapping disks. For example, in Figure 2.5, an object X is divided into subobjects $X0, X1, ..., Xn$. Then, $X0$ is stored in cluster 0, $X1$ is stored in cluster 1 and so on.
- Then, a subobject is divided into fragments. The fragments of a subobject are striped across the disks *within* a cluster so that consecutive fragments of subobject $X0$ (say, $X0.i$ and $X0.i+1$) are stored in consecutive disks within a cluster. For example, subobject $X0$ in turn consists of fragments

$X0.0, X0.1, X0.2$ and $X0.3$. Then, fragment $X0.0$ is stored in disk 0 (of cluster 0), $X0.1$ is stored in disk 1 (of cluster 0) and so on.

Hence, while retrieving the object X, the server will use cluster C_0 first, then switch to cluster C_1, and then to C_2, and then the cycle repeats. Every time the server switches to a new cluster, it incurs an overhead in terms of the disk seek time. Taking this switching overhead time (say t_{switch}) into account, the server can schedule object retrieval from the next cluster t_{switch} time ahead of its *normal* schedule time. This simple data striping works better for media objects with similar data transfer rate requirements. This is because the disks are divided into fixed number of clusters and the server schedules the cluster operations in the same sequence. The disadvantage of this approach is that striping objects with different data retrieval rate requirements becomes difficult.

Staggered Striping Technique : In order to provide better performance for media objects with different data transfer rate requirements, staggered striping technique can be used. Here, disks are not divided into clusters but treated as independent storage units. An object is striped in a staggered manner as follows :

- The first fragment of consecutive subobjects are located at a distance of k disks where k is termed the *stride*. Figure 2.6 shows an assignment for a media object X using staggered striping technique with stride $k = 1$. Here, the first fragment $X0.0$ is located in disk 0 and $X1.0$ in disk 1.
- The consecutive fragments of the same object are stored in successive disks. In Figure 2.6, the fragments of the object X, $X0.0$ is stored in disk 0, $X0.1$ in disk 1 and $X0.2$ in disk 3.

The advantage of staggered striping is that media objects with different data transfer rate requirements can easily be accommodated by choosing different values for the stride (k). As an example, text data requires lower bandwidth and hence can be stored with a higher value of stride. Video data requires higher bandwidth and hence can be stored with lower value of stride.

Network Striping : Striping of multimedia objects can be carried out across disks connected to a network as shown in Figure 2.7. Each multimedia server has a cluster of disks and the entire group of clusters is managed in a distributed manner. The data can be striped using standard or staggered (or any other) striping technique. Network striping assumes that the underlying

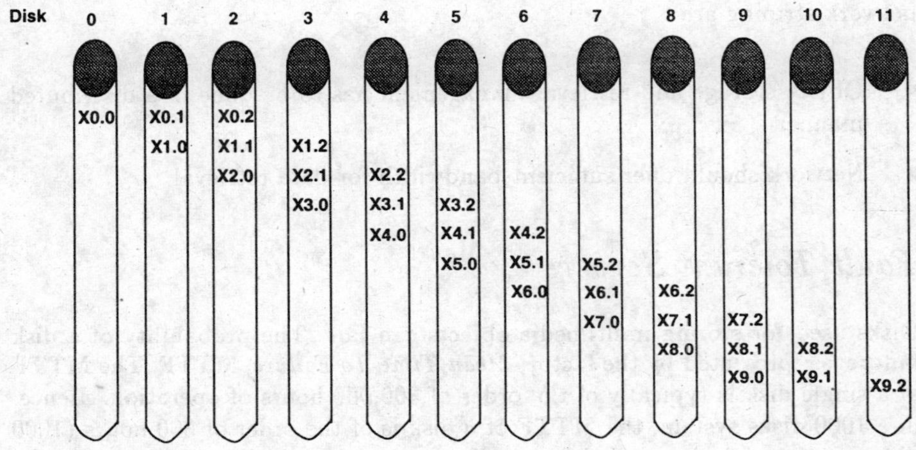

Figure 2.6 Staggered Striping Technique

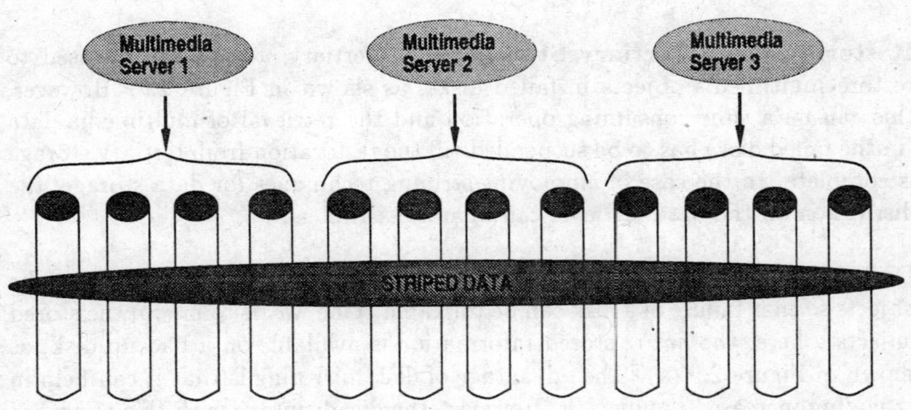

Figure 2.7 Network Striping Technique

network has the capability to carry data at the required data transfer rate. Network striping helps in improving data storage capacity of multimedia systems and also helps in improving data transfer rates. The disadvantages of network striping are :

- Object storage and retrieval management has to be done in a distributed manner
- Network should offer sufficient bandwidth for data retrieval

Fault Tolerant Servers

Disks used for storing multimedia objects can fail. The probability of a disk failure is represented by the factor, *Mean Time To Failure*, MTTF. The MTTF of a single disk is typically of the order of 300,000 hours of operation. Hence, in a 1000 disks system, the MTTF of *a* disk is of the order of 300 hours (1000 disks server might be needed for applications such as VoD). In the event of a failure, the following strategies can be used :

- Restoration from tertiary storage
- Mirroring of disks
- Employing parity schemes

Restoring From Tertiary Storage : Tertiary storage can be used to restore multimedia objects in failed disks, as shown in Figure 2.8. However, this can be a time consuming operation and the retrieval of multimedia data (in the failed disk) has to be suspended till the restoration from tertiary storage is complete. In the case of employing striping techniques for data storage, the disruption on the data retrieval can be quite significant.

Disk Mirrors : A better alternative is to store some redundant multimedia objects so that failure of a disk can be tolerated. One way is to *mirror* the stored objects : here, the *entire* stored information is available on a backup disk, as shown in Figure 2.9 (a). The advantage of disk mirroring is that it can help in providing increased bandwidth. However, the disadvantage with this approach is that it might become very costly in terms of the required disk space.

Employing Parity Schemes : Another alternative is to employ parity schemes, as shown in Figure 2.9 (b). Here, an object is assumed to be striped

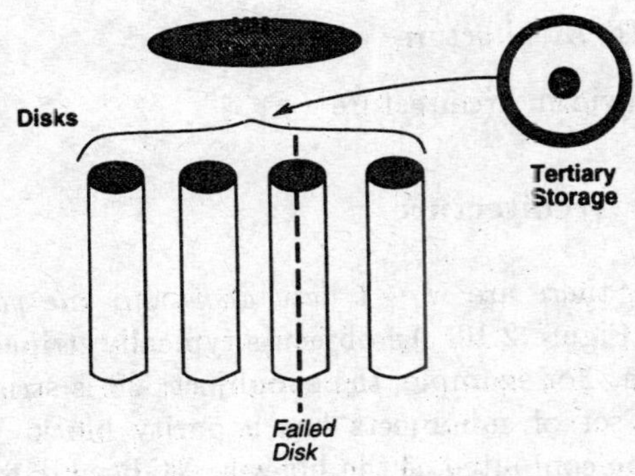

Figure 2.8 Using Tertiary Storage As Backup

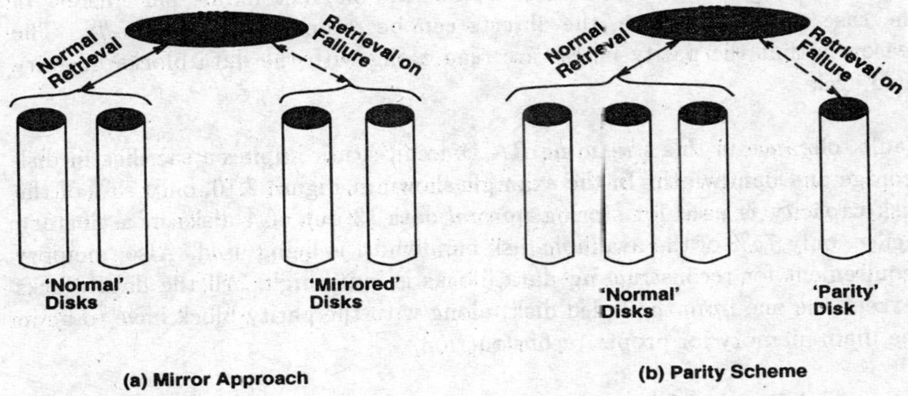

Figure 2.9 Fault Tolerant Servers

across three disks and the fourth stores the parity information. In the case of failure of *a data* disk, the information can be restored by using the parity information.

In the event of a disk failure, the lost data can be restored by using the parity fragment along with the fragments from normal disks. For reconstruction of the lost data, all the object fragments have to be available in the buffer. Also, the disk used for storing parity block cannot be overloaded with *normal* object fragments. This is because at the time of failure of a disk the retrieval of parity blocks might have to compete with that of the normal fragments. The following strategies can be adopted for storing the parity information.

- Streaming RAID Architecture
- Improved Bandwidth Architecture

Streaming RAID Architecture :

In this architecture, there are $N-1$ *data disks* and one *parity disk* for each cluster, as shown in Figure 2.10. An object is typically striped over all the data disks, as data blocks. For example, the subobject $X0$ is striped as $X0.0, X0.1$ and $X0.2$ and this set of subobjects has a parity block $X0.p$. The *parity fragment* $X0.p$ can be computed as the bit-wise XOR-ed data of the fragments $X0.0, X0.1$ and $X0.2$: $X0.p = X0.0 \oplus X0.1 \oplus X0.2$. The sequence of subobjects $(X0, X1, ..)$ are then striped across the clusters, as in the case of simple striping. The streaming RAID architecture can tolerate one disk failure per cluster. In the case of a disk failure, the objects can be reconstructed *on-the-fly*. The reason is that the parity blocks are read along with the data blocks in every read cycle.

Fault tolerance in this streaming RAID architecture implies a sacrifice in disk storage and bandwidth. In the example shown in Figure 2.10, only 75% of the disk capacity is used for storing *normal* data (3 out of 4 disks in a cluster). Hence, only 75% of the available disk bandwidth is being used. Also, memory requirement for reconstructing data blocks is quite high. All the data blocks (except the one from the failed disk) along with the parity block have to be in the main memory for proper reconstruction.

Improved Bandwidth Architecture :

An improvement that can be made is on the disk bandwidth utilization. When parity blocks are stored on separate disks, one disk is sacrificed during normal operations. Instead, data and parity blocks can be inter-mixed to improve the disk bandwidth, by storing the parity block of disk cluster i in the cluster $i+1$. Such an improved bandwidth architecture is shown in Figure 2.11. During normal read operations, parity blocks are not scheduled for reading. When a disk failure occurs, the parity block in the cluster $i+1$ is scheduled for reading and the missing data is reconstructed. The advantage here is that no separate disk is dedicated as a parity disk, leading to an improvement in bandwidth. The disadvantage is that reading of parity blocks in a cluster has to be scheduled along with other data blocks. This can result in overloading of disk(s) in a cluster. In the case where disk bandwidth is not sufficient to allow for both data and parity blocks, the cluster can drop some data blocks giving priority to the parity blocks.

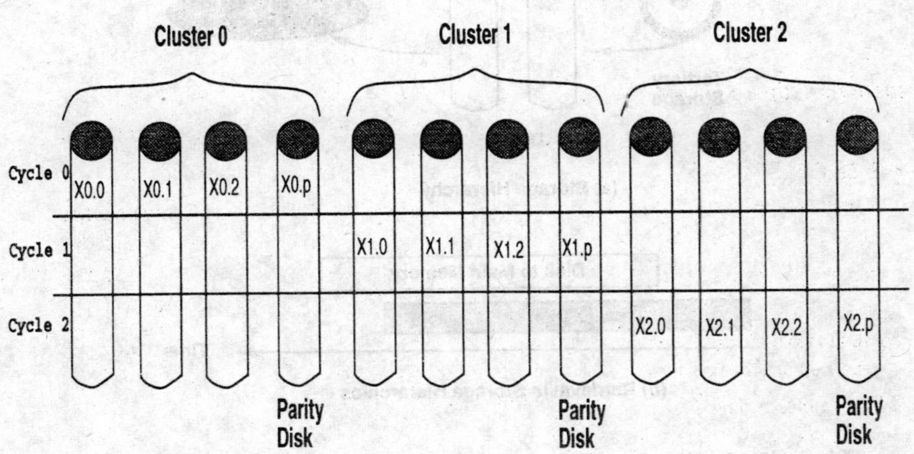

Figure 2.10 Streaming RAID Architecture

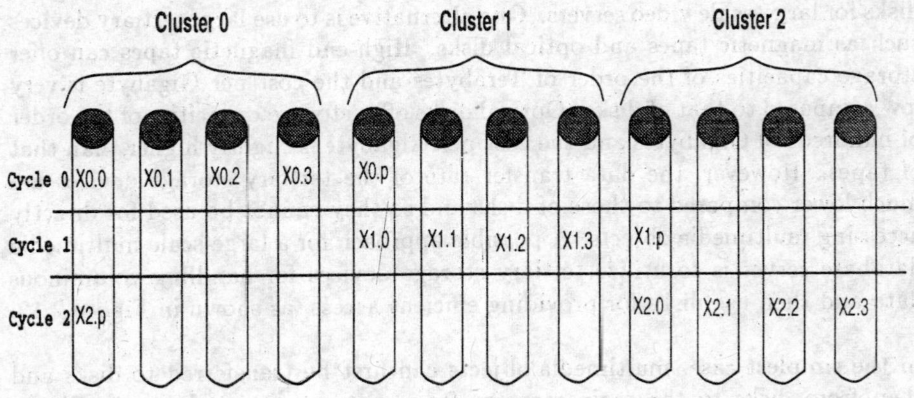

Figure 2.11 Improved Bandwidth Architecture

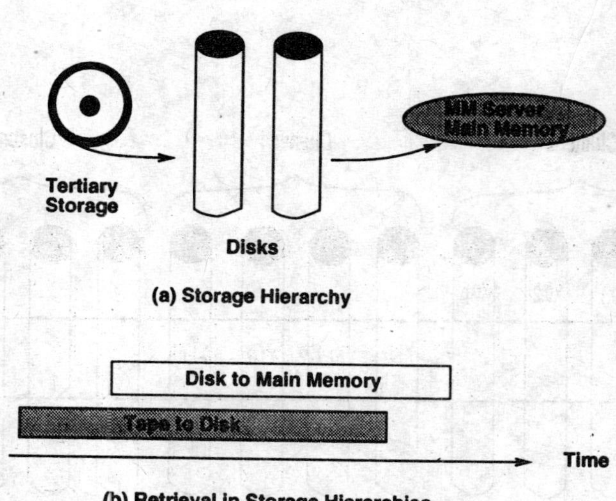

Figure 2.12 **Storage Hierarchy in A Multimedia Server**

2.1.3 Utilizing Storage Hierarchies

The discussions we have had so far focussed on effective storage of objects on disks. Disks do provide efficient means for storage and retrieval of multimedia objects. However, they are limited by the storage capacity that can be offered (it is in terms of Gigabytes) and the cost per Gigabyte inhibits the usage of disks for large-scale video servers. One alternative is to use large tertiary devices such as magnetic tapes and optical disks. High-end magnetic tapes can offer storage capacities of the order of Terabytes and the cost per Gigabyte is very low compared to that of disks. Optical disks offer storage capacities of the order of hundreds of Gigabytes and the cost per Gigabyte is slightly higher than that of tapes. However, the data transfer rate of the tertiary storage devices are much lower compared to those of disks and so they cannot be used for directly accessing multimedia objects. A possible approach for a large-scale multimedia database server is to utilize tertiary storage devices for handling voluminous data and then use disks for providing efficient access, as shown in Figure 2.12.

In the simplest case, multimedia objects can first be transferred to disks and then from disks to the main memory for consumption, as shown in Figure 2.12(b) (consumption of objects can be by their display or by communicating them to clients). Object transfers from tapes to disks is necessary because the data transfer rates of tertiary storage devices cannot match the consumption rates of objects such as video. This approach leads to longer initial wait times before the objects can be made available for consumption. It is possible to reduce initial wait times by storing initial portions of objects in disks.

2.1.4 Comparison of Storage Techniques

As discussed above, we have different schemes for storing multimedia objects. The selection of a particular scheme for object storage in a multimedia database server depends on the following factors.

- Number of simultaneous multimedia applications that need to be supported
- Bandwidth required by each application
- Support for VCR like functions, such as fast forward and rewind
- Amount of required storage
- Reliability considerations, including their susceptibility to catastrophic failures and degradation of service
- Affordable cost

Based on the above factors, one needs to compare different storage schemes and select an appropriate one for the multimedia database server. Depending on the amount of storage and the required application bandwidth, we need to select the number and type of disks. Once that is done, we need to select the type of striping technique that can be used to store the data. The striping technique can be selected based on the characteristics of the disks : the number of available disks, bandwidth offered by the disks, seek times and rotational latencies of the disks, etc. In the case of network striping, we also need to consider the characteristics of the network such as available network throughput.

Similarly, in the case of need for fault tolerance, we need to determine the following factors.

- **Disk storage :** The amount of disk storage that is needed to ensure fault tolerance, e.g., the size of the parity information.
- **Bandwidth :** The amount of bandwidth that must be dedicated to ensure fault tolerance, e.g., the bandwidth needed for transferring parity information. This bandwidth cannot be used for transferring normal data.

Type of Storage	Techniques Used
Single Disk & Multiple Disks with Media on a Disk	(a) Continuos Storage (b) Randomly Scattered Storage (c) Constrained Storage (d) Merged Storage (e) Log-structured Storage
Multiple Disks with Striping	(a) Simple Striping (b) Staggered Striping (c) Network Striping
Fault Tolerance	(a) With Tertiary Storage (b) Disk Mirrors (c) Parity Schemes : (i) Streaming RAID (ii) Improved Bandwidth Architecture

Table 2.1 Multimedia Storage Techniques

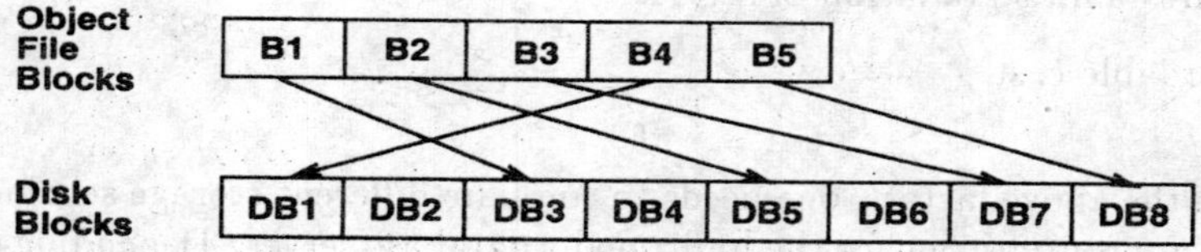

Figure 2.13 Association Between File and Disk Blocks

- **Buffer space :** The amount of memory required for buffering data in order to reconstruct the missing data.

Table 2.1 summarizes the various techniques discussed for multimedia storage.

2.2 FILE RETRIEVAL STRUCTURES

Till now, we discussed how object blocks are allocated and stored on disks. Another important issue in the design of multimedia data storage is to keep track of the association between disk blocks and multimedia objects (or files). For instance, Figure 2.13 shows an example of the association between file and

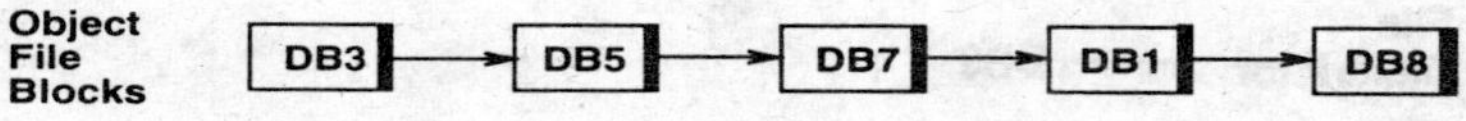

Figure 2.14 Disk Blocks Links

disk blocks. Here, object block $B1$ is stored in disk block $DB3$, $B2$ in $DB5$, and so on. We need mechanisms to map object blocks to disk blocks so that they can help in :

- Travelling from one disk block to another in a fast manner
- Accessing multimedia objects in a random manner

The following techniques can be used to store the disk block to multimedia object blocks association.

Linked Disk Blocks : Here, the end of each disk block contains a pointer to the next block in the file. The file descriptor only needs to store the pointer to the first block. This is a simple solution but random access to multimedia data implies accessing *all* the previous data blocks. Figure 2.14 shows the linked disk blocks for the multimedia object storage shown in Figure 2.13.

File Allocation Table (FAT) : Here, a file descriptor contains an entry to the first block, as shown in the Figure 2.15 (for object storage shown in Figure 2.13). Then, a table (FAT) is used where an entry for *each disk block* maintains its successor disk block. The FAT in Figure 2.15 shows the successor of each disk block starting from $DB1$. An empty successor entry indicates that a disk block has no link to another block. Continuous access to objects can be done by starting from the block pointed by the file descriptor ($DB3$ in this example) and using the FAT entries to find the successors ($DB5, DB7, DB1$ and $DB8$, in this example). Random access can be made by accessing the FAT directly. However, considering the amount of disk space that can be associated with a multimedia database server, the FAT can turn out to be very huge.

File Index : The FAT approach discussed above maintains the information for the entire disk. Instead, each object can have an index that describes the ordered list of disk blocks associated with that file. There is no need to maintain a separate file allocation table. Figure 2.16 describes this index approach for object storage shown in Figure 2.13. Random access can be made by walking through the disk blocks list. This index information has to be stored in the disk

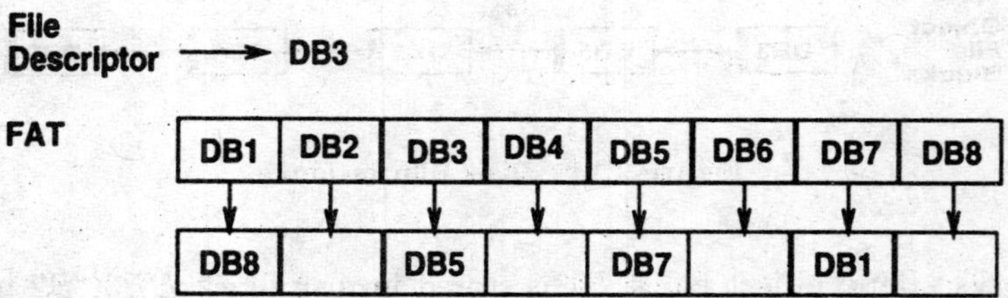

Figure 2.15 File Allocation Table

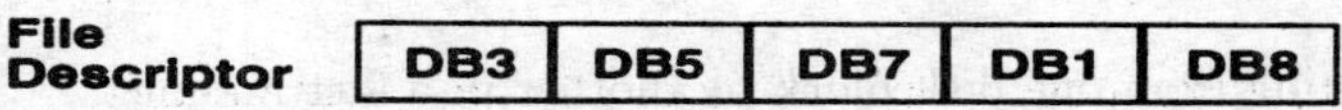

Figure 2.16 File Index

like another object. The disadvantage with this approach is that multimedia servers might need to keep a number of large files open. In this case, the number of indexes that have to be maintained in the memory increases linearly.

Hybrid Approach : In order to provide efficient continuous as well as random access, we can employ a hybrid approach. For continuous access to multimedia objects, we can employ linked disk blocks. For random access, we can download the index corresponding to the accessed file.

2.2.1 Summary

File retrieval structures maintain the association between object blocks and disk blocks. Since multimedia object retrieval may be done in continuous or random manner, file retrieval structures need to support both. We discussed techniques such as linked disk blocks, FAT, file index, and hybrid approach. Hybrid approach, with linked disk blocks for continuos access and indexes for random access, seems to be a better choice for multimedia objects. Table 2.2 summarizes the techniques used for file retrieval structures.

2.3 DISK SCHEDULING

The discussion so far was with respect to storage of objects in multimedia database servers. During normal operations, multimedia database servers re-

File Retrieval Structure	Techniques Used
	(a) Linked Disk Blocks (b) File Allocation Table (c) File Index (d) Hybrid Approach

Table 2.2 Techniques For File Retrieval

ceive a large number of data retrieval requests. These requests might involve high volumes of data transfer with real-time constraint for delivering blocks of data in periodic intervals. Hence, these requests may have to be processed over multiple read cycles. The methodology adopted for scheduling the read requests influence the real-time data requirements of the multimedia applications. The following algorithms are used for scheduling the read requests :

- Earliest Deadline First (EDF)
- Round Robin
- Disk Scan
- Scan-EDF
- Grouped Sweep Scheme

Earliest Deadline First : This is the best known algorithm for real-time scheduling of tasks with deadlines. As the name indicates, this scheduling algorithm processes requests with earliest deadlines for retrieval. The disadvantage of the EDF algorithm is that it results in poor server resource utilization. This is because successive requests might involve random disk accesses, resulting in excessive seek times and rotational latencies.

Round Robin : This scheme processes requests in *rounds*, with the multimedia server retrieving atmost one data block for each application request in each round. In the round-robin scheme, the order in which the read requests are processed is fixed across the rounds. As shown in Figure 2.17 (a), a read request scheduled first in round i is scheduled first in round $i + 1$ also. This results in the maximum time between successive retrievals for a request being bounded by a round's duration. The advantage of the round robin scheme is that there is no need for extra buffering of data to satisfy the real-time data transfer requirements. The disadvantage is the same as that of the EDF scheme, it may result in excessive seek times and rotational latencies.

Disk Scan : Here, requests are optimized from the server point of view by scheduling the tasks with shortest disk seek times first. This methodology helps in improving the disk throughput. However, the disadvantage is that the real-time constraints of a read request may not always be satisfied since the seek time of the request need not be the shortest. As shown in Figure 2.17 (b), a request scheduled first in round i might be scheduled last in round $i+1$.

Scan-EDF : This algorithm combines the Scan technique with EDF. Scan-EDF processes requests with the earliest deadlines first, just like the EDF. However, when several requests have the same deadline, the requests are processed based on the shortest seek time first, just like the Scan scheme. The effectiveness of the Scan-EDF method depends on the number of requests having the same deadline.

Grouped Sweep Scheme : This scheme basically helps in *grouping* or *batching* a set of requests. Each round typically consists of a set of groups of requests. *Within* the group, the Scan scheduling scheme is applied by processing requests with the shortest seek time first. The groups themselves are serviced in round robin. As shown in Figure 2.17 (c), a request scheduled first in group $G1$ of round i can be scheduled last in the group $G1$ of round $i+1$. Hence, the maximum time between reads is bound by the duration of the round and the maximum group read time.

2.3.1 Summary

Read access to multimedia objects might involve transfer of information over a period of time. The transfer of information involves real-time constraints for delivering blocks of data in periodic time intervals. Hence, disk access requests have to be scheduled appropriately. Table 2.3 summarizes the techniques discussed for disk scheduling.

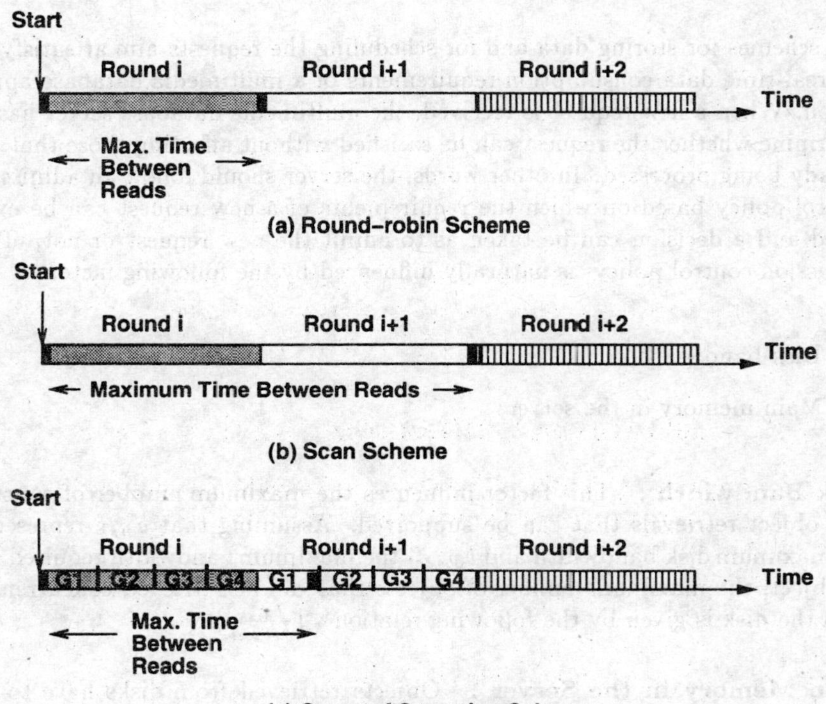

Figure 2.17 Disk Scheduling Schemes

Disk Scheduling	Techniques Used
	(a) Earliest Deadline First (EDF) (b) Round-robin (c) Scan (d) Scan-EDF (e) Grouped Sweep Scheme

Table 2.3 Techniques For Disk Scheduling

2.4 SERVER ADMISSION CONTROL

The schemes for storing data and for scheduling the requests aim at satisfying the real-time data consumption requirements of a multimedia database application. When a new request is received, the multimedia database server has to determine whether the request can be satisfied without affecting those that are already being processed. In other words, the server should follow an admission control policy based on which the requirements of a new request can be evaluated and a decision can be taken as to admit the new request or not. This admission control policy is naturally influenced by the following factors :

- Disk bandwidth
- Main memory in the server

Disk Bandwidth : This factor influences the maximum number of concurrent object retrievals that can be supported. Assuming that b_{disk} represents the maximum disk bandwidth and b_{object} the maximum bandwidth required for an object, the maximum number of objects that can be retrieved concurrently from the disk is given by the following relation : $\lfloor \frac{b_{disk}}{b_{object}} \rfloor$.

Main Memory in the Server : Objects retrieved from disks have to be held in the main memory of the server before they are consumed (i.e., either displayed or communicated to the client). In order to make a simple estimate of the required main memory in the server, let us make the following assumptions.

- Let an object be divided into n equi-sized subobjects with each subobject requiring B bytes.
- Let C denote the consumption rate of the object.
- Let T_{disk} denote the time required for retrieval from disks and $T_{consume}$ denote the consumption time ($T_{consume} = \frac{B}{C}$).

Figure 2.18 shows the memory requirement for concurrent retrieval of four objects, assuming that the objects are similar in nature (i.e., the size of the subobjects and the consumption rate are same). Consider the memory requirement of each object at a time instant t_1 : subobject $O1_1$ requires no memory, $O2_1$ requires $B/3$ memory, $O3_1$ requires $2B/3$ memory, and $O4_1$ requires B memory. Hence, the total memory requirement for concurrent retrieval of these four

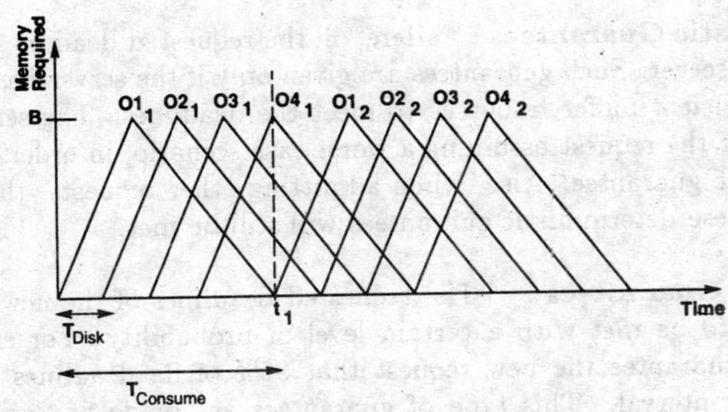

Figure 2.18 Memory Requirement for Four Concurrent Retrievals

objects is $2B$. It has been proved that, for concurrent retrieval of N objects (with each subobject requiring B bytes), the total memory required is $\frac{NB}{2}$. (Interested readers can refer [132] for a complete proof). Hence, if a multimedia database server with a main memory of Mem needs to support N concurrent object retrievals, then the following constraint must be satisfied : $\frac{NB}{2} \leq Mem$.

2.4.1 Admission of New Requests

While admitting a new request, the real-time requirements of the request have to be evaluated. For example, some applications can tolerate missed deadlines: couple of lost video frames per minute or a couple of seconds silence in audio. The server might be able to admit such a request with a degree of tolerance towards failed deadlines, even under high loads. Hence, for admitting a new request, the multimedia database server should :

- Evaluate the worst-case seek time and rotational latencies of the disks
- Evaluate the requirements of the requests that are already being processed
- Evaluate the real-time requirements of the new request and its tolerance towards missed deadlines

After making the above evaluations, the multimedia database server has to check whether the disk bandwidth and the main memory requirements are satisfied. Then, it can offer the following guarantees to the new request.

Deterministic Guarantees : Here, all the requested deadlines are guaranteed by the server. Such guarantees are given only if the server has a light load and has sufficient buffer resources to meet the deadlines. The server reserves resources for the request assuming a worst case scenario, in order to provide a deterministic guarantee. Also, while admitting other requests, the server has to ensure these deterministic guarantees will still be met.

Statistical Guarantees : The requested deadlines of the new request are guaranteed to be met with a certain level of probability. For example, the server can guarantee the new request that 95% of its deadlines will be met over a time interval. This type of guarantees are made by considering the statistical behavior of the system as well as the tolerance levels specified by the new request. While admitting another request, the server has to ensure that the guaranteed level of statistical service can still be maintained for the earlier requests. In the instances where deadlines have to be missed, the server has to ensure that the same request does not get penalized repeatedly.

No Guarantees (Background Processing) : Here, no guarantees are provided by the multimedia server. Requests are scheduled only when the server has time left after scheduling all the deterministic and statistical ones.

Server Admission	Types of Guarantees
	(a) Deterministic Guarantee (b) Statistical Guarantee (c) Background

Table 2.4 Server Admission Control

2.4.2 Summary

The policies used for storing multimedia objects, for storing file retrieval structures, for scheduling disk access requests, can all be effective only when a multimedia database server admits a fixed number of requests based on their requirements. Each request necessitates certain data transfer rate from the multimedia server over a period of time. Hence, before admitting a new request, a server has to identify whether it can satisfy the request's requirements as well as those of the other requests currently being serviced. Table 2.4 summarizes the types of guarantees a server can provide to the requests.

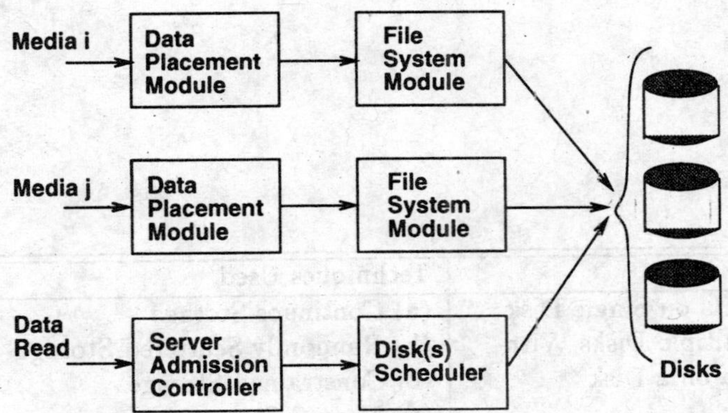

Figure 2.19 Components of Storage Manager

2.5 CONCLUDING REMARKS

Storage and retrieval of multimedia objects are characterized by high volume of information and real-time requirements. High volume of information necessitates large amounts of disk space. Real-time requirements imply that the disk bandwidth should be sufficient to store and retrieve data at the rate required by the application. In this chapter, we discussed the schemes that can be used for storing multimedia data. We also discussed fault tolerant schemes that can be used for improving the reliability of multimedia database servers.

For satisfying the real-time requirements for data storage and retrieval, we considered the policies that can be employed for scheduling disk access requests. Based on the schemes employed for storage and disk access scheduling, we discussed how multimedia database servers can guarantee disk access service by controlling the admissions to the server. Table 2.5 summarizes the techniques for multimedia storage and retrieval discussed in this chapter.

Based on the discussions we have had so far, we can visualize a simple storage manager as shown in Figure 2.19. The data placement module determines how objects belonging to different media types can be stored. The file system module updates the file retrieval structures associated with the objects to be stored. For handling a new read request, the server admission control module determines whether the requirements of the read request can be satisfied. If the requirements can be satisfied, then the disk(s) scheduler module schedules the request.

Issue	Techniques Used
Objects on Single Disk & Multiple Disks With Media on a Disk	(a) Continuos Storage (b) Randomly Scattered Storage (c) Constrained Storage (d) Merged Storage (e) Log-structured Storage
Objects on multiple Disks with Striping	(a) Simple Striping (b) Staggered Striping (c) Network Striping
Fault Tolerance	(a) With Tertiary Storage (b) Disk Mirrors (c) Parity Schemes : (i) Streaming RAID (ii) Improved Bandwidth Architecture
File Retrieval Structure	(a) Linked Disk Blocks (b) File Allocation Table (c) File Index (d) Hybrid Approach
Disk Scheduling	(a) Earliest Deadline First (EDF) (b) Round-robin (c) Scan (d) Scan-EDF (e) Grouped Sweep Scheme
Server Admission	(a) Deterministic Guarantee (b) Statistical Guarantee (c) Background

Table 2.5 Multimedia Storage Techniques

Bibliographic Notes

An overview of multimedia storage servers can be found in [140]. Issues in designing a multi-user HDTV storage server are discussed in [90]. A multimedia filing system has been described in [24, 30].

Constrained disk block allocation has been introduced in [99, 140]. Different techniques for merging objects with constrained gaps between successive pair of data blocks has been discussed in [99]. A matrix-based disk allocation strategy for low-cost VoD servers has been proposed in [154, 161]

Redundant Array of Inexpensive Disks (RAID) architecture was proposed in [27]. The technique of staggered striping technique was introduced by [118]. It also provides a performance comparison study of simple and staggered striping architectures, and concludes that the staggered striping technique performs better. [137] provides a detailed discussion on fault tolerant design of multimedia servers. It also gives a comparison of various fault tolerant schemes based on factors such as disk space overhead, bandwidth overhead, and mean time to failure.

Utilization of memory and disk bandwidth has been discussed in [132]. Server admission control mechanisms and the guarantees that can be offered to disk access requests are examined in [140].

3

METADATA FOR MULTIMEDIA

Media objects such as audio, image, and video are binary, unstructured and hence un-interpreted. Multimedia databases have to derive and store interpretations based on the content of these media objects. These interpretations are termed *metadata*. Metadata is *data about data*. The metadata has to be generated automatically or semi-automatically (or in some cases manually) from the media information.

3.1 METADATA : CLASSIFICATION

Metadata generation can be based on either intra-media or inter-media information. The intra-media metadata deals with the interpretation of information *within* the media. On the other hand, the inter-media metadata deals with the interpretation of information on multiple media and their relationships. Figure 3.1 describes the ways metadata can be generated from media objects. Here, the extracting functions $f1, f2, ..., fn$ work on the individual media and generate the intra-media data $i1, i2, ..., in$. The extracting function F generates inter-media metadata, $I1, I2, ..., In$, by working on all the composed media. The functions applied to extract metadata can be dependent or independent of the contents of media objects. Based on the type of extracting function used, metadata can be classified into the following categories :

- Content-dependent metadata
- Content-descriptive metadata
- Content-independent metadata

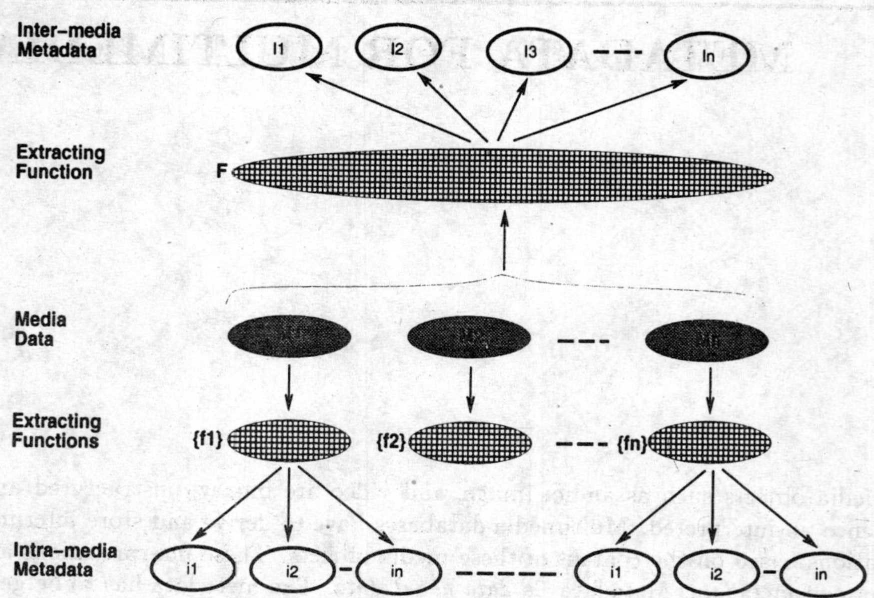

Figure 3.1 Intra-media and Inter-media Metadata

Content-dependent Metadata : depends only on the content of media objects. Derivation of facial features of a person's photographic image (such as the type of nose or ear, color of hair) and derivation of camera operations (such as panning, tilting and zooming) in a video clip belong to this category.

Content-descriptive Metadata : is associated with the media information, but cannot be generated automatically from their contents alone. This type of metadata describes the characteristics of media objects with support of user's or application's impressions. For example, metadata on facial expression such as anger or happiness depends on the contents of the image but has to be derived with support from users or tools which can support such cognitive process.

Content-independent Metadata : does not depend on the contents of the media information but is associated with it. Name of the photographer who took the picture, budget of a movie, author(s) who created a multimedia document are examples of content-independent metadata.

3.1.1 Metadata Generation Methodologies

The generation of metadata is done by applying feature extracting functions on media objects. As discussed before, these feature extracting functions employ content-dependent, content-independent, and content-descriptive techniques. These techniques use a set of terminologies that represents an application's view as well as the contents of media objects. These set of terminologies, referred to as *ontologies*, create a semantic space onto which metadata is mapped. The ontologies used for derivation of metadata are :

Media Dependent Ontologies : These ontologies refer to the concepts and relationships that are specific to a particular media type, such as text, image, or video. For example, features such as color or texture can be applied to image data whereas features such as silence periods can be applied only to audio data.

Media Independent Ontologies : These ontologies describe the characteristics that are independent of the contents of media objects. For example, the ontologies corresponding to the time of creation, location, owner of a media object are media independent.

Metacorrelations : Metadata associated with different media objects have to be correlated to provide a unified picture of a multimedia database. This unified picture, called the *query metadata*, is modeled for the needs of database users based on application-dependent ontologies. As an example, consider the following query on geographic information database on the Himalayan mountain ranges : *Get me images of the peaks which are atleast 20,000 feet in height, showing atleast 5 people climbing the peaks.* Here, a correlation has to be established between the height of the peaks in the Himalayas and their available images.

Figure 3.2 shows the various stages in the process of generation of metadata. The media pre-processor helps in *identifying* the *contents of interest* in the media information. The *contents of interest* can be, for example, a word in a spoken sentence or a shot in a movie clip. For each media, a separate pre-processor is required to identify the contents of interest. Then, different ontologies are used to interpret the contents of interest. In the following sections, we discuss the metadata as well as the pre-processing techniques for different media.

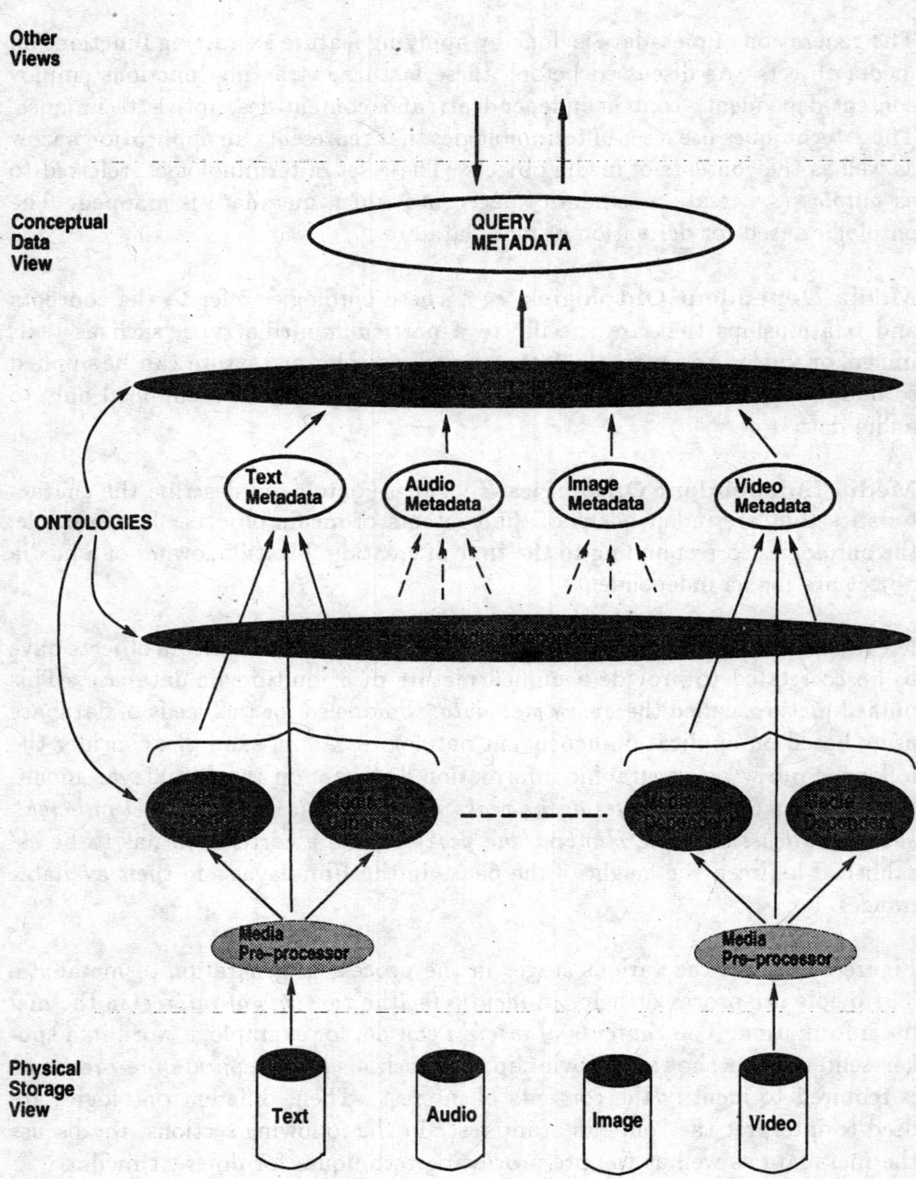

Figure 3.2 Ontologies and Metadata Generation

3.2 METADATA FOR TEXT

Text is often represented as a string of characters, stored in formats such as ASCII. Text can also be stored as digitized images, scanned from the original text document. Text has a logical structure based on the type of information it contains. For example, if the text is a book, the information can be logically structured into chapters, sections, subsections, glossary, index, etc. Metadata for text describes this logical structure, and the generation of metadata for text involves identifying its logical structure. Text that is keyed into a computer system normally uses a language to represent the logical structure. In the case of scanned text images, mechanisms are needed for identifying columns, paragraphs, semantic information and for locating the key words. The semantic information, the detected keywords, the located columns and paragraphs then serve as the metadata for text images.

3.2.1 Types of Text Metadata

The following types of metadata can be used to describe text objects.

Document Content Description : This metadata provides additional information about the content of the document. As an example, a document on global warming can be described as a document in environmental sciences.

Representation of Text Data : This metadata describes the format, coding and compression techniques that is used to store the data. For example, the language using which the text has been formatted is a metadata. This metadata is content-descriptive.

Document History : This metadata describes the history of creation of the document. This metadata is also content-descriptive. The components described as part of the history could include :

- Status of the document : for example, draft, technical report, conference paper, journal paper, etc.
- Date of update of the document including the author(s) who did the update
- Components of the older document which have been modified

Document Location : This content-independent metadata describes the location (workstation or a personal computer) in a computer network that holds the document.

3.2.2 Generating Text Metadata

Metadata for text can be described by the author(s) using text formatting languages. Standard Generalized Markup Language (SGML) is a language specified by the International Standards Organization (ISO) to help in describing typographical annotations as well as the structure of the document. (*Markup* is a notion from the publishing world, where manuscripts are annotated with a pencil or a marker to indicate the layout). However, the markups or the structure provided by the author(s) alone may not be sufficient in many instances. Hence, automatic or semi-automatic mechanisms might be required to identify metadata such as topic and subtopic boundaries. Such mechanisms are especially needed when text is present in the form of scanned images. Here, we discuss the following text metadata generation methodologies :

- Text formatting language : SGML
- Automatic/semi-automatic mechanisms :
  (a) Subtopic Boundary Location
  (b) Word-Image Spotting

Text Formatting Language

SGML is a text structuring language and is an ISO standard from 1986. SGML is a markup language. The general methodology for marking up is by using dedicated symbols or characters. The disadvantage of this approach is the *contamination* of the contents of the document by the structure and layout commands. SGML segregates the contents of the document and its layout information. Documents in SGML are composed of basic building blocks, termed *elements*. Document components such as title, figures, paragraphs, and footnotes are SGML elements. The possible ordering of the SGML elements and their relationships describe the document structure. These structures conform to *Document Type Definitions* (DTD). The DTD specifies how element types can be built from other element types, i.e., it specifies the logical structure of a document. An element in the DTD specification is described by its name and the way that an element of that type can be constructed. For example, a DTD element of type *JournalPaper* can be composed of *TitleInfo, Abstract, Contents,* and *References.* The element definition of *JournalPaper* in DTD will then be :

```
<!ELEMENT JournalPaper - - (TitleInfo, Abstract, Contents,
References)>
<!ELEMENT TitleInfo - - (Authors, Affiliation, Address)>
<!ELEMENT Contents - - (Sections)>
<!ELEMENT Sections - - (Paragraphs, Figures, Tables)>
```

All the elements in a document should be completely defined in DTD. Additional properties of an SGML element can be described by means of *attributes.* Attributes help in expressing the characteristics of elements (and hence of the documents). For example, the attribute list of the element *JournalPaper* can include its date of publication, details of publication such as volume, number and the title of the journal, etc. The attributes of an SGML element, *JournalPaper*, are defined in DTD as follows :

```
<!ATTLIST   JournalPaper
            date_of_publication   DATE                 #REQUIRED
            volume                ID1                  #REQUIRED
            number                ID2                  #REQUIRED
            journal_title         ID3                  #REQUIRED
            availability          (available | no)     available>
```

Here, the name of the element type for which the attributes are defined, is given immediately after the keyword **ATTLIST**. Each attribute is defined with a name, type of the attribute (*date_of_publication* belongs to the type *DATE*), followed by an optional default value or an optional directive. In the above example, the attribute *availability* has the default value *available.* The directive for handling the attribute is a preceded by the #-symbol. For example, the directive *REQUIRED* indicates that a value for the attribute has to be specified.

The DTD specifies an ordered tree or a *parse* tree, of the elements composing the document. The vertex of the tree is the SGML element and the edge of the tree defines *partOf-* relationship. Figure 3.3 shows the tree structure of the *JournalPaper* DTD.

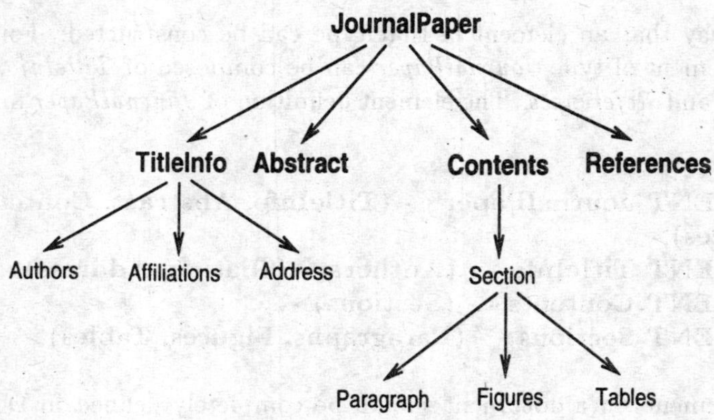

Figure 3.3 DTD Tree For Journal Paper

Metadata from SGML Specification : The DTD definition of a SGML document is a metadata that describes the structure of the document. The mapping from the document components to element information is also part of the metadata. The attributes defined as part of the element definition serves as metadata.

Automatic/Semiautomatic Mechanisms

Metadata that is derived from text formatting languages such as SGML, is those that are declared by the author(s) of the document. This metadata may or may not reflect all the semantic aspects of the document. One might need to use automatic/semi automatic mechanisms to generate metadata dealing with other semantic aspects of the document. Here, we discuss two such mechanisms: subtopic boundary location and word-image spotting.

Subtopic Boundary Location : *TextTiling* algorithms are used for the purpose of partitioning text information into *tiles* that reflects the underlying topic structure. The basic principle in the TextTiling algorithm is that terms describing a subtopic co-occurs locally and a switch to another subtopic implies co-occurrence of a different set of terms. The algorithm identifies subtopic boundaries by :

- Dividing or *tokenizing* the text into 20-word adjacent token sequences. In TextTiling, a block of *k* sentences (the value of *k* being determined by heuristics) is treated as a logical unit.

- Comparing the adjacent blocks of token-sequences for overall lexical similarity. The frequency of occurrence of a term within each block is compared to its frequency in the entire domain. This helps in identifying the usage of the term within a discussed topic or in the entire text. If the term occurs frequently over the entire text, then it cannot be used to identify topics. On the other hand, if the occurrence frequency is localized to a block or a set of co-occuring blocks, it can be used to identify topics in the text.
- Computing similarity values for adjacent blocks. Determining boundary changes by changes in the sequence of similarity scores.

Word-Image Spotting : In the case of digitized text images, keywords have to be *located* in the document. The set of keywords that are to be located, can be specified as part of the application. Typical word-spotting systems need to do the following :

- Identify a text line by using a bounding box of a standard height and width. The concept of *multi-resolution morphology* is used to identify text lines using the specified bounding boxes. Interested readers can refer [63] for a detailed discussion on this.
- Identify specific words within the (now) determined text line. A technique termed, *Hidden Markov Model* (HMM), is used to identify the specific words in the text line. HMMs are described in Section 3.3.1.

3.2.3 Summary

Though text can be considered as the *simplest* of media objects (in terms of storage requirements, representation, ease of identification of the content information, etc.), it is very heavily used to convey the information. It forms an integral part of multimedia database applications and plays a vital role in representation and retrieval of information. Text can be represented as a string of characters (using ASCII) or as a digitized image. In the case of text being represented as a string of characters, we need a language to describe the logical structure. We discussed the features of SGML for describing the logical structure of text. In many instances, the description provided by a language may not be sufficient to identify the content information. Hence, we need automatic mechanisms to identify topics and keywords in the text. Also, in the case of text images, we need to identify the keywords that occur in the text images. Towards this purpose, we discussed automatic mechanisms for helping

Text Representation	Issues	Mechanisms
ASCII String	Description of logical structure	Languages like SGML
	Topic identification	Algorithms like TextTiling
Digitized Images	Keyword Spotting	HMM models

Table 3.1 Text Metadata Generation

in identifying topic boundaries and in identifying occurrence of keywords in text images. Table 3.1 summarizes the issues in metadata generation for text.

3.3 METADATA FOR SPEECH

The speech media refers to the spoken language and is often not *defined* as an independent data type. It is considered as part of audio. The importance of speech processing arises due to its ease of use as input/output mechanism for multimedia applications. The metadata that needs to be generated can be content-dependent or content-descriptive. The metadata generated for speech can be as follows.

- Identification of the spoken words. This is called *speech recognition*, and helps in deciding whether or not a particular speaker produced the utterance. It is also termed *verification*.
- Identification of the speaker. Here, a person's identity is chosen from a set of known speakers. It is called *speaker identification* or *speaker recognition*.
- Identification of prosodic information which can be used for drawing attention to a phrase or a sentence, or to alter the word meaning.

Metadata generated as part of speech recognition is content-dependent. This metadata can consist of the start and the end time of the speech, along with a confidence level of the spoken word identification. Metadata generated as part of speaker recognition can be considered as content-descriptive, though this metadata is generated by analyzing the contents of the speech. This metadata can consist of the name of the speaker, the start and the end time of the speech. Metadata describing the prosodic information can be considered as content-dependent. It can consist of the implied meaning in case the speaker altered the word meaning and a confidence score of the recognition of the prosodic information. Content-independent metadata can also be associated with speech data. Time of the speech, location where the speech was given, format in which speech data is stored, can be considered as content-independent metadata for speech. In addition, silence periods and non-speech sounds can be identified and stored as metadata.

3.3.1 Generating Speech Metadata

The process of speech and speaker recognition is very complex. The most general form of recognition is the one where there is no limitation either on the vocabulary (called *text-independent*) or on the number of speakers, is still very inaccurate. However, the recognition rates can be made high by controlling the vocabulary as well as the number of speakers. The following five factors that can be used to control and simplify the task of speech and speaker recognition.

1. **Isolated words :** Isolated words are much easier to recognize than continuous speech. The reason is that isolated words have silence periods in between, which serve as word boundaries. The coarticulation effects in continuous speech cause the pronunciation of words to be modified, depending on its position relative to other words in a sentence. This leads to difficulties in recognizing the word boundaries.

2. **Single Speaker :** The parametric representations of speech are highly sensitive to the characteristics of the speakers. This makes a recognition system to work better for a single speaker.

3. **Vocabulary Size :** Similar to the number of speakers, the size of the vocabulary to be recognized also plays an important role. The probability of having similar sounding words in a larger vocabulary is much higher than in a small vocabulary.

4. **Grammar :** For spoken sentence recognition, the allowed sequence of words plays an important role. The allowable sequence of words is called the grammar of the recognition domain. A tightly constrained grammar allows only a limited set of words to follow any given word and helps better in speech recognition.

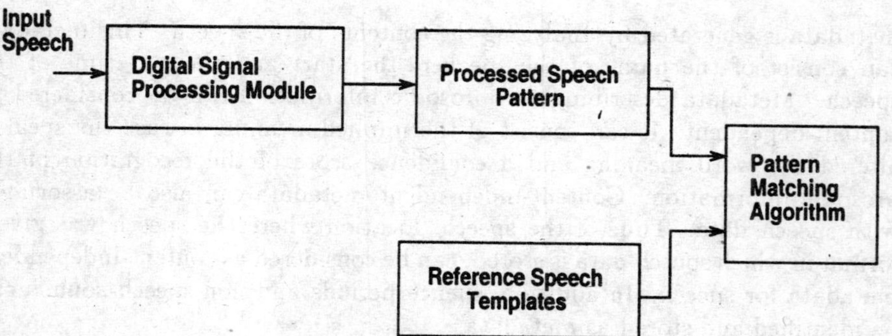

Figure 3.4 Speech Recognition System

5. **Environment :** The environment in which the speech to be recognized is produced, influences the accuracy of recognition. The environmental characteristics include the background noise, changes in microphone characteristics and loudness. However, it is not always possible to control the environment where speech is produced.

Now, we shall describe the components of a possible speech recognition system.

Speech Recognition System

A typical speech recognition system has two main components as shown in Figure 3.4 :

- Signal processing module
- Pattern matching module

The signal processing module gets the speech analog signal (through a microphone or a recorder), and digitizes it. The digitized signal is processed to do the following actions : detection of silence periods, separation of speech from non-speech components, conversion of the raw waveform into a frequency domain representation and data compression. The stream of such sample speech data values is grouped into frames of usually 10 - 30 milliseconds duration. The aim of this conversion is to retain only those components that are useful for recognition purposes.

This processed speech signal is used for identification of the spoken words or the speaker or prosodic information. The identification is done by matching the processed speech with stored patterns. The pattern matching module has a repository of reference patterns that consists of the following :

- Different utterances of the same set of words (for speech recognition)
- Different utterances by the same speaker (for speaker verification)
- Different ways of modifying the meaning of a word (for identifying prosodic information)

Pattern Matching Algorithms

For recognition, the speech data to be recognized has to be compared with the stored training templates or models. This necessitates algorithms to compute a measure of similarity between the template(s) and the sample(s). The following algorithms are popular for speech recognition :

- Dynamic Time Warping
- Hidden Markov Models (HMM)
- Artificial Neural Networks models

Dynamic Time Warping : The comparison of the speech sample with the template is conceptually simple if the preprocessed speech waveform is compared directly against a reference template, by summing the distances between respective speech frames. The summation provides an overall distance measure of similarity. The simplicity of this approach is complicated by the non-linear variations in timing produced from utterance to utterance. This results in misalignment of the frames of the spoken word with those in the reference template. The template can be stretched or compressed at appropriate places to find an optimum match. This process of time "warping" on the template to find the optimum match is termed *Dynamic Time Warping*. Dynamic programming procedure can be used to find the best warp that minimizes the sum of distances in the template comparison. Figure 3.5 shows the use of Dynamic Time Warping to help in speech pattern matching.

Hidden Markov Models (HMM): HMMs have underlying stochastic finite state machines (FSMs). The stochastic state models are defined by the following.

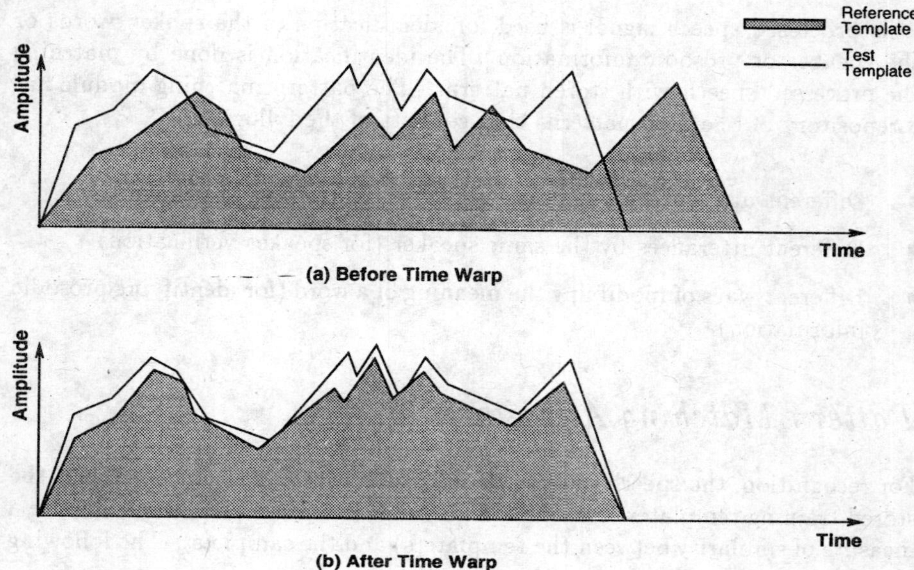

Figure 3.5 Dynamic Time Warp : An Example

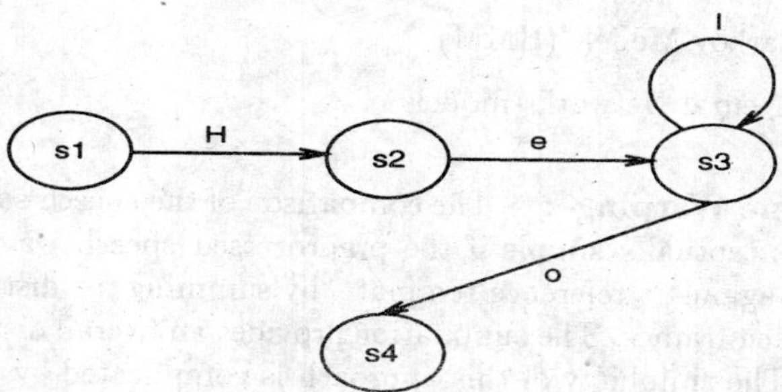

Figure 3.6 Hidden Markov Model : An Example

- A set of states
- An output alphabet
- A set of transition and output probabilities

A HMM for word recognition is constructed with a template having a set of states, with the arcs between any two states representing a positive transition probability, as shown in Figure 3.6. Here, $\{s1, s2, s3, s4\}$ are the set of states. The output alphabets are {H,e,l,o}. The HMM in this example is designed to recognize the word : *Hello*. The transition probabilities are defined between each pair of states. The output probabilities are associated with each transition, defining the probability of emitting each output alphabet while a particular transition is made. The example in Figure 3.6 does not show the transition and output probabilities. The term ***hidden*** for this model is due to the fact that the actual state of the FSM cannot be observed directly, only through the alphabets emitted. Hence, a hidden Markov model can be considered as one that generates random sequences according to a distribution determined by the transition and output probabilities. The probability distribution can be discrete or continuous. For isolated word recognition, each word in the vocabulary has a corresponding HMM. For continuous speech recognition, the HMM represents the domain grammar. This grammar HMM is constructed from word-model HMMs.

HMMs have to be ***trained*** to recognize isolated words or continuous speech. The process of training involves setting the probabilities involved so as to increase the probability of a HMM generating the desired output sequences. The given set of output sequences are the training data. The following algorithms associated with the HMMs for the purpose of training :

- Forward algorithm for recognizing isolated words
- Viterbi algorithm for recognition of continuous speech

The function of the forward algorithm is to compute the probability that a HMM generates an output sequence. A sequence of processed speech codes is recognized as a certain word if the probability that the corresponding HMM generates this sequence is maximal. The forward algorithm is used in isolated word recognition. The Viterbi algorithm determines the state transition path, based on the grammar model for the continuous speech to be recognized. The word models are connected in parallel for recognizing the continuously spoken words.

Artificial Neural Networks Models : Artificial neural networks (ANNs) is an information processing system that simulates the cognitive process of the human brain. The basic idea is to build a ***neural structure*** that can be ***trained*** to perform the cognitive function of the input signals. The neural network consists of a number of very simple and highly interconnected processors termed ***neurodes***, which are analogous to the neurons in the brain. These neurodes are

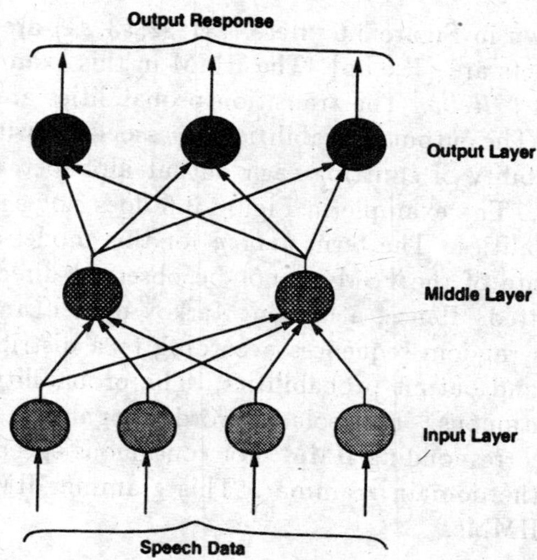

Figure 3.7 Artificial Neural Network : Typical Architecture

connected by a large number of links that have weighted functions associated with them. The neurodes *communicate* their *decisions* among themselves over the weighted links. The decision of a neurode might be given different weights for different links.

Figure 3.7 shows a typical architecture of an artificial neural network. The neural network is organized as a layered architecture. The input layer of neurodes receive the input data. The decision of the neurodes in the input layer is conveyed to the neurodes in the middle layer through the weighted links. The neurodes in the middle layer can receive inputs from more than one neurodes in the input layer. The middle layer neurodes convey their decisions to those in the output layer. In practice, the middle layer can be absent or can be comprised of more than one layer.

In order to determine the weights for the links connecting the neurodes, the neural network has to be *trained.* The training procedure consists of presenting the input data such as speech templates and describing the desired output. During this training process, the neural network *learns* how to recognize the input data. During this learning process, the link weights are assigned.

Requirements for speech metadata	Mechanisms Discussed
Analog-to-digital conversion of speech signal	Digital Signal Processing techniques
Identification of speech, speaker and prosodic speech	Pattern matching algorithms : Dynamic time warping, Hidden Markov model, Artificial Neural Networks

Table 3.2 Speech Metadata Generation

Prosodic Speech Detection

Emphatic speech is characterized by the modification of pitch, volume, and timing. The speaking volume is estimated by computing the energy in a short duration of a speech signal. Features that can be used for prosodic speech detection include the fundamental frequency, energy changes in the fundamental frequency, and the energy in the speech signal. HMM models of different prosodic patterns are used for comparing the prosodic features derived from a speech signal. The temporal information of the detected prosodic information can then be used as a metadata.

3.3.2 Summary

Speech provides a very flexible medium for input and output to multimedia database applications. Some security features for the applications can be implemented using speaker identification mechanisms. Generation of speech metadata requires identification of the spoken words/sentences, the speaker, and the prosodic (or emphatic) speech. We discussed the methodologies used for identifying these metadata. Table 3.2 summarizes the issues in metadata generation for speech.

3.4 METADATA FOR IMAGES

Metadata for images depend on the type of images that are to be analyzed and the application(s) that will be using the analyzed data. We consider the metadata that can be used for a few types of images such as satellite images, facial images, and architectural design images.

Metadata for Satellite Images : The satellite images, as viewed by computer scientists, are treated as three-dimensional grids (a regular grid), with 2889 rows, 4587 columns and 10 layers deep. The perception of earth scientists is to focus on the processes that created the images. From this point of view, the image has 10 *bands* or *layers*, each created by a different process. The following broad categories of metadata can be defined for the satellite metadata.

- *Raster metadata* : describes the grid structure (rows, columns and depth of the grid), spatial, and temporal information. The spatial information describes the geographic coordinates (latitudes and longitudes) and overlay of the image on another (with a state or county boundary, for example). The temporal information describes the time at which the image was taken.
- *Lineage metadata* : includes the processing history : algorithms and parameters, used to produce the image.
- *Data set metadata* : describes the sets of data available at a particular site as well as the detailed information about each data set.
- *Object description metadata* : includes the structure of a database table or the specific properties of an attribute (for example, the data type of an attribute such as latitude or longitude).

Metadata for Architectural Design : Architectural design deals with design, cost estimation, and 3D visualization of buildings. The following metadata for architectural design can be identified. Rooms in a building, number of windows and doors, ceiling heights, and floor area are content-dependent metadata. The location of a building and its address are content-descriptive metadata. Architect's name, company name, and cost estimate are content-independent metadata.

Metadata for Facial Images : The content-dependent metadata are the facial features of humans such as color of hair, description of eyes, nose, and mouth. The content-descriptive metadata can include sex and race. Name of a person, social security number and other details form the content-independent metadata.

3.4.1 Generating Image Metadata

Algorithms used for generating the required metadata are better off when they know the type of images that are being analyzed. The algorithms can then use specific information on the properties of the image type for taking decisions. For example, algorithms used for generating metadata for satellite images need not worry about the relative locations of left and right eyes on a human face. Hence, algorithms for feature analysis are unique depending on the type of images that are being analyzed. Apart from feature extraction, one might need to analyze the color and texture information of the objects as well. The following steps are involved in extracting the features from images.

- **Object locator design :** The basic requirement in image feature extraction is to locate the objects that occur in an image. This requires the image to be segmented into regions or objects. Designing the object locator is to select an image segmentation algorithm that can isolate individual objects.

- **Feature selection :** The specific properties or the features of objects are to be determined in this step. These features should help in distinguishing different types of objects that might occur in the set of images that are to be analyzed.

- **Classifier design :** This step helps in establishing the mathematical basis for determining how objects can be distinguished based on their features.

- **Classifier training :** The various adjustable parameters (such as the threshold values) in the object classifier must be fixed so as to help in classifying objects. Design and training of the classifier module are specific to the type of images. For instance, classifiers for architectural design are different from those used for satellite images.

Image Segmentation

The process of image segmentation helps in isolating objects in a digitized image. There are two approaches to isolate objects in an image. One approach, called the *boundary detection approach*, attempts to locate the boundaries that exist among the objects. Other approach, called the *region approach*, proceeds by determining whether pixels fall inside or outside an object, thereby partitioning the image into sets of interior and exterior points. We shall describe few techniques that can be used in image segmentation.

Thresholding Technique : The principle behind this technique is that all pixels with gray level at or above a threshold are assigned to object. Pixels below the threshold fall outside the object. This technique falls under the region approach and helps in easy identification of objects in a contrasting background. Determination of the value of the threshold has to be done carefully since it influences the boundary position as well as the overall size of the object.

Region Growing Technique : This technique proceeds as though the interior of the object grows until their borders correspond with the edges of the objects. Here, an image is divided into a set of tiny regions which may be single pixel or a set of pixels. Properties that distinguish the objects (such as gray levels, color or texture) are identified and a value for these properties are assigned for each region. Then, the boundary between adjacent regions is examined by comparing the assigned values for each of the properties. If the difference is below a certain value, then the boundary of the two regions is dissolved. This region merging process is continued till no boundaries can be dissolved.

Storing Segmented Image

Different techniques are used to store the identified objects *within* an image as well as their spatial characteristics. These techniques can also help in identifying intersection of the objects in the image. Section 4.3.1 carries a detailed discussion on this topic.

Feature Recognition, An Example : Facial Features

After segmentation of an image, the objects in the image have to be classified according to the desired features. This involves the steps : feature selection, classifier design, and classifier training, as discussed in Section 3.4.1. These steps depend on the type of image whose objects are to be classified as well as the application. For instance, the features to be selected and the object classifier for satellite images will be different from the ones for facial images. We shall describe briefly how the segmented image can be used for extracting facial features.

For facial feature recognition, the objects to be identified include left eye, right eye, nose, mouth, ears, etc. The area of search for a particular object can be reduced by applying the relationships between objects known *apriori*. For example, in a facial image, we know that mouth is *below* the nose, right eye

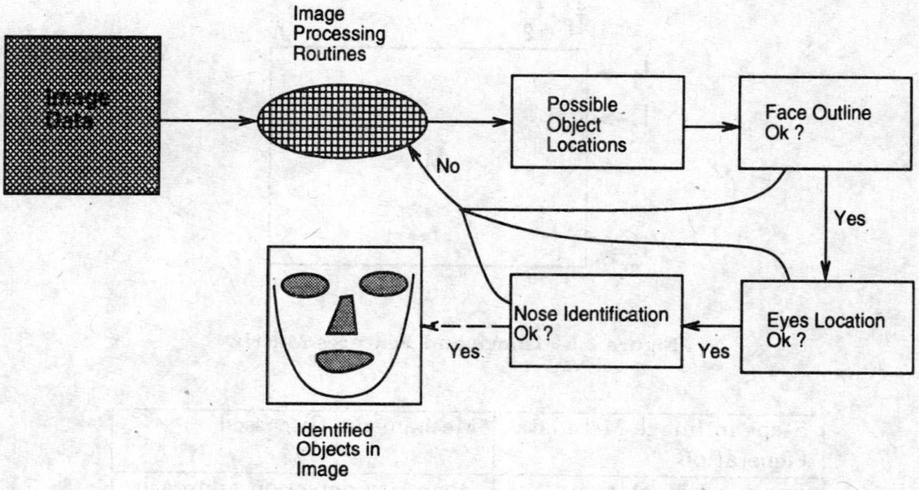

Figure 3.8 Steps in Facial Recognition

should be at a distance *d* from the left eye and so on. Figure 3.8 shows the steps involved in the feature extraction from a facial image. The first step is to determine the face outline. Once the outline is detected, eyeballs can be located. When one eye ball is located, the other can be located within a distance. Then, nose can be identified with the constraint that the bottom of the nose should be between the horizontal centers of the eye balls, and approximately half the vertical distance from eyes to the chin. A score of certainty is also specified with each extracted feature. In case the certainty score is low, alternate mechanisms can be used. These mechanisms include using a relaxed facial template, re-examining a previously located feature (in case the present feature depends on it) or getting user's input.

Mathematical Model for Using Metadata

The generated metadata has to be represented in an easily accessible manner. Metadata can be represented by a m x n matrix, M. As shown in Figure 3.9, this matrix has m image objects $(i_i, i = 1, ...m)$. Then, each image is represented by the n-dimensional feature distribution $(f_1, ..., f_n)$. If a metadata feature corresponds to the image, it is given a value 1.0. Otherwise, it is given a value 0.0. If the feature works in a negative manner, it is given a value -1.0. This matrix gives a metadata space that can be used as the search space for extracting images when a user query describing image features is given.

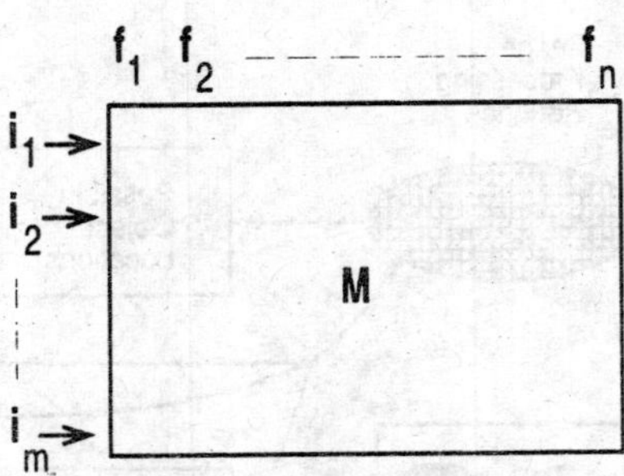

Figure 3.9 Image and Features Matrix

Steps in Image Metadata Generation	Mechanisms Discussed
Image Segmentation	Boundary detection approach, region growing approach
Storing Segmented Image	Discussed in Section 4.3.1
Feature Identification	Depends on the type of image and application

Table 3.3 Image Metadata Generation

3.4.2 Summary

Metadata for images involves identification of the objects that are present in an image. In this section, we described how images can be segmented into composing objects and how these objects can be classified according to a set of desired features. Table 3.3 summarizes the issues involved in generating image metadata.

3.5 METADATA FOR VIDEO

Video is stored as a sequence of frames, by applying compression mechanisms to reduce the storage space requirements. The stored data is raw or un-interpreted in nature and hence interpretations have to be drawn from this raw data. The metadata on video can be on : (i) sequence of video frames, (ii) a single video frame. The following video metadata can be identified :

Content-dependent : This type of metadata describes the *raw* features of video information. For a sequence of video frames, this metadata can include camera motion (such as pan, tilt), camera height, lighting levels, and the track of objects in the video sequence. At individual frame level, the metadata can describe frame characteristics such as color histograms. (Color histograms store extracted color features in the form of histograms with the histogram value indicating the percentage of pixels that are most similar to a particular color). In a similar manner, gray level sums and gray level histograms can be used to describe grayscaled images.

Content-descriptive : For a sequence of video frames, this metadata can consist of features such as camera shot distance (close-up, long, medium), shot angle, shot motion, action description, type of objects in the shot, etc. For a single frame, the metadata can consist of features such as frame brightness, color, texture, type of objects in the frame, description of objects, etc.

Content-independent : This metadata describes features that are applicable perhaps to a whole video, instead of a sequence of frames for a smaller interval. The description may consist of features such as production date, producer's name, director's name, budget of the video, etc.

3.5.1 Generating Video Metadata

The easiest form of generating metadata for video is to provide textual descriptions. These descriptions may be manually *logged* and stored as associated database information. Alternatively, automatic/semi-automatic mechanisms can be used to generate the required metadata. The content-dependent metadata features can be extracted by applying algorithms for automatic partitioning of video data. The content-descriptive metadata generation uses application-dependent ontologies to describe the contents of video objects. The content-independent metadata has to be generated based on the inputs given about a video object by a user or an application. To help in the process of generating the video metadata, the tools should have the following functions :

- Identify logical information units in the video

- Identify different types of video camera operations
- Identify the low-level image properties of the video (such as lighting)
- Identify the semantic properties of the parsed logical unit
- Identify objects and their properties (such as object motion) in the video frames

The logical unit of information that is to be parsed automatically is termed a *camera shot* or a *clip*. A shot is assumed to be a sequence of frames representing a contiguous action in time and space. The basic idea behind the identification of shots is that the frames on either side of a camera break shows a significant change in the information content. Algorithm used in the video parser should be able to detect this change in the information content, and hence identify the shot boundaries. The algorithm needs a quantitative metric that can capture the information content of a frame. Based on the fact whether the difference between the metrics of two consecutive video frames exceed a threshold, a shot boundary can be identified by the algorithm. This idea for identifying camera shots gets complex when fancy video presentation techniques such as dissolve, wipe, fade-in or fade-out are used. In such cases, boundary between two shots no longer lies between two consecutive frames, instead is spread over a sequence of frames.

Two types of metrics are used to quantify and compare the information content of a video frame :

- Comparison of corresponding pixels or blocks in the frame
- Comparison of histograms based on color or gray-level intensities

The available video information may be compressed or uncompressed. Hence, the video parsing algorithm might have to work on compressed or uncompressed information.

Algorithms for Uncompressed Video

These algorithms work on uncompressed video, implying that for a compressed source, the information has to be uncompressed before it can be analyzed.

Histogram Based Algorithm : The extracted color features of a video frame are stored in the form of *color bins* with the histogram value indicating

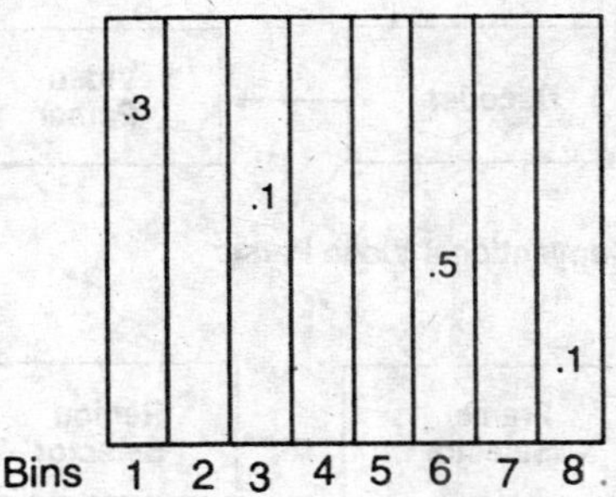

Figure 3.10 Two Dimensional Color Histogram

the percentage (or the normalized population) of pixels that are most similar to a particular color. Each bin is typically a cube in the 3-dimensional color space (corresponding to the basic colors red, green, and blue). Any two points in the same bin represent the same color. A typical color histogram with eight bins is shown in Figure 3.10. Similarly, gray levels in black and white images can also be stored in the form of histograms.

Video shot boundaries can be identified by comparing the following features between two video frames : gray level sums, gray level histograms, and color histograms. In this approach, video frames are partitioned into sixteen windows and the corresponding windows in two frames are compared based on the above features. This division of frames helps in reducing errors due to object motion or camera movements. This approach does not consider gradual transition between shots. For overcoming this short-coming, two different levels of thresholds can be adopted : one for camera breaks and the other for gradual transitions.

Algorithms for Compressed Video

Compressed video can be in motion JPEG, MPEG, or other formats. Different techniques have been developed for parsing compressed video. These techniques use the features of the specific compression methods for parsing the video data.

For Motion JPEG Video : JPEG compression standard applies to color as well as gray scaled images. Motion JPEG is a fast coding and decoding technique that can be applied to video frames. In motion JPEG, a video frame is grouped into data units of 8 * 8 pixels and a Discrete Cosine Transform

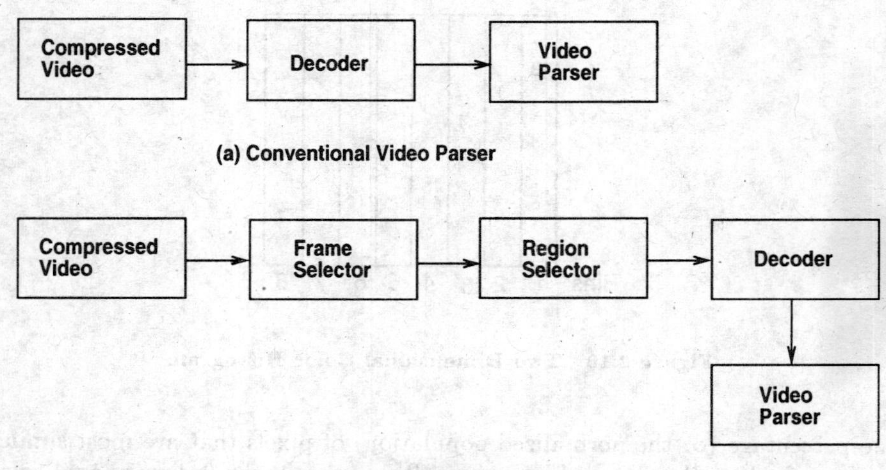

Figure 3.11 Selective Decoder for Motion JPEG Video

(DCT) is applied to these data units. The DCT coefficients of each frame are mathematically related to the spatial domain and hence represents the contents of the frames. Video shots in motion JPEG can be identified based on correlation between the DCT coefficients of video frames. The identification of shot boundaries is done in two stages :

- Apply a skip factor to select the video frames to be compared
- Select *regions* in the selected frames. Decompress only the selected regions for further comparison

Figure 3.11(b) shows the block diagram of the motion JPEG video parser. The frame selector uses a *skip factor* to determine the subsequent number of frames to be compared. The region selector employs a DCT coefficients based approach to identify the regions for decompression and for subsequent image processing. The algorithm adopts a multi-pass approach with the first approach isolating the regions of potential cut points. Then, the frames that cannot be classified based on DCT coefficients comparison are decompressed for further examination by color histogram approach.

A conventional video parser decodes *all* the frames and parses the frames based on the comparison between the histograms, as shown in Figure 3.11(a). On the other hand, the selective decoding technique helps in reducing the overheads involved in decompressing *all* the frames before their comparison. The disadvantages with the selective decoding approach are that it does not help in detecting shot boundaries in the presence of gradual transitions, camera operations, and object motions.

For MPEG Video : MPEG standard aims at compressing the video so that the data rate is about 1.2 Mbits/s. MPEG compresses video frames in the following manner.

- To achieve high rate of compression, redundant information in the subsequent frames are coded based on the information in the previous frames. Such frames are termed **P** and **B** frames.
- To provide fast random access, some of the frames are compressed independently. Such frames are called **I** frames.

I frames (*Intra coded frames*) are self-coded, i.e., coded without any reference to other images. An **I** frame is treated as a still image and hence compressed using JPEG. **P** frames (*Predictive coded frames*) are compressed with respect to the information in the previous **I** and **P** frames. **B** frames (*Bi-directionally predictive coded frames*) are used for reverse presentation of video frames. They are compressed based on the previous **I** and **P** frames. Hence, we can consider a MPEG video stream to be of the following sequence of frames : *IBBPBBPBBIBBPBBP....*

Parsing MPEG coded video source can be done by using the following metrics.

- A difference metric for comparison of DCT coefficients between video frames is used. The difference metric using the DCT coefficients can however be applied only on the **I** frames of the MPEG video, since those are the only frames that are coded with DCT coefficients.
- Motion information coded in the MPEG data can be used for parsing. The basic idea here is that in MPEG, the **B** and **P** frames are coded with motion vectors, and the residual error after motion compensation is transformed and coded with DCT coefficients. The residual error rates are likely to be very high at shot boundaries. Hence, the number of motion vectors in the **B** or **P** frame is likely to be very few. So the algorithm detects a shot boundary if the number of motion vectors are lower than a threshold value.

This approach can lead to detection of false boundaries because a shot boundary can lie between two successive **I** frames. The advantage is that the processing overhead is reduced as the number of **I** frames are relatively fewer. The algorithm also does partitioning of the video frames based on motion vectors.

For detecting shot boundaries in the presence of gradual transitions, a hybrid approach of employing both the DCT coefficient based comparison and motion vector based comparison. The first step is to apply a DCT comparison to the **I** frames with a large skip factor to detect regions of potential gradual transitions. In the second pass, the comparison is repeated with a smaller skip factor to identify shot boundaries that may lie in between. Then the motion vector based comparison is applied as another pass on the **B** and **P** frames of sequences containing potential breaks and transitions. This helps in refining and confirming the shot boundaries detected by DCT comparisons.

Detection of Camera Operations and Object Motions

Camera operations and object motions induce a specific pattern in the field of motion vectors. Panning and tilting (horizontal or vertical rotation) of the camera causes the presence of strong motion vectors corresponding to the direction of the camera movement. The disadvantage of using this idea for detection of pan and tilt operations is that movement of a large object or a group of objects in the same direction can also result in a similar pattern for the motion vectors. To distinguish object movements from camera operations, the motion field of each frame can be divided into a number of macro blocks and then motion analysis can be applied to each block. If the direction of all the macro blocks agree, it is considered as arising due to camera operation (pan/tilt). Otherwise it is considered as arising due to object motion. In zoom operation, a focus center for motion vectors is created, resulting in the top and bottom vertical components of motion vectors with opposite signs. Similarly, the leftmost and the rightmost horizontal components of the motion vectors will have the opposite sign. This information is used for identification of zooming operation.

3.5.2 Summary

Video has to be processed for extracting the required metadata. This processing involves detection of video shots, object motions and camera movements. We discussed techniques that help in doing these for both uncompressed and compressed video. Table 3.4 summarizes the issues in video metadata generation.

Video Representation	Issues	Mechanism
Uncompressed	Shot detection	Histogram based models, Production based model
Motion JPEG	Shot detection	DCT coefficients based approach
MPEG	Shot detection, camera operations and objects movement	Hybrid approach (DCT coefficients for **I** frames; Motion vectors metric for **B** & **P** frames)

Table 3.4 Video Metadata Generation

3.6 CONCLUDING REMARKS

Metadata is basically data about data. Data belonging to media such as text, speech, image, and video are either unstructured or partially structured. Interpretations, based on the contents of media objects as well as on an application, have to be *derived* from the raw data. Based on how the interpretations are derived, metadata is classified as content-dependent, content-descriptive, and content-independent. A set of terminologies, termed *ontologies*, that reflect the application's view of the information as well as the contents of the media information, are used for deriving the required metadata. For the ontologies to work on the contents of the media information, pre-processing techniques have to be used to *extract* the contents. We discussed some pre-processing techniques used for different types of media information. Table 3.5 summarizes the issues and the mechanisms used for generating metadata.

Figure 3.12 shows a simple block diagram of metadata manager that does the function of generating and maintaining the metadata associated with the media objects in the database. The media pre-processor module identifies the contents of interest in different media objects. These contents of interest are classified according to the set of ontologies used and the metadata for the media objects are generated. The metacorrelations module correlates the various media metadata and generates the query metadata. Updates to the generated metadata can either be in the form of modifications to the media objects or to the set of ontologies used.

Media Type	Media Representation	Issues	Mechanisms Discussed
Text	ASCII String	Description of logical structure	Languages like SGML
		Topic identification	Algorithms like TextTiling
	Digitized Images	Keyword Spotting	HMM Models
Speech		Analog-to-digital conversion of speech signal	Digital Signal Processing techniques
		Identification of speech, speaker & prosodic speech	Pattern matching : Dynamic time warping, HMM
Image		Image Segmentation	Boundary detection, region growing, etc.
		Storing Segmented Image	Discussed in Section 4.3.1
		Feature Identification	Depends on image and application
Video	Uncompressed	Shot detection	Histogram comparison
	Motion JPEG	Shot detection	DCT coefficients based approach
	MPEG	Shot detection, camera operations, objects movement	Hybrid approach

Table 3.5 Metadata Generation For Different Media

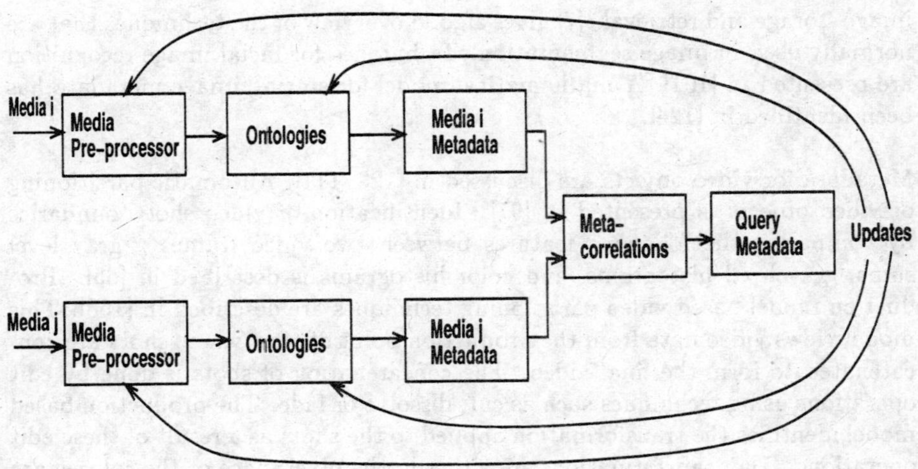

Figure 3.12 Components of Metadata Manager

Bibliographic Notes

Issues in generation of metadata for multimedia objects have been discussed in [121, 126]. The strategies for application and media dependent metadata derivation are described in [157]. It also provides a classification of the ontologies used for deriving multimedia metadata.

[122] describes different types of metadata for text. Text structuring language, SGML, has been introduced in [23, 143]. *TextTiling* algorithms have been proposed for the purpose of partitioning text information into *tiles* that reflects the underlying topic structure [87, 88, 128, 129]. Several word-spotting systems have been proposed in the literature [128, 95]. The concept of ***multi-resolution morphology***, used to identify the text lines using the specified bounding boxes, has been discussed in [63]. Hidden Markov Models (HMM) has been introduced in [33].

Metadata for speech has been described in [128]. [37] identifies the factors that can be used to control and simplify the task of speech and speaker recognition. HMM for speech metadata generation has been introduced in [83, 127]. Neural networks model for speech recognition has been described in [86, 131].

Metadata for satellite images are described in [125]. Metadata for architectural design are identified in [149]. [73] describes the metadata requirements for facial image storage and retrieval. [7] gives a good overview of the techniques that are normally used in image segmentation. Techniques for facial image recognition are presented in [101]. A mathematical model for storing image metadata has been identified in [124].

Metadata for video objects are discussed in [123, 111]. Automatic partitioning of video objects is presented in [97]. Identification of video shot boundaries by comparing the following features between two video frames : gray level sums, gray level histograms, and color histograms is described in [65]. Production model based video partitioning techniques are described in [158]. This model views video data from the production point of view where shots are concatenated to form the final video. The concatenation of shots is done by edit operations using techniques such as cut, dissolve or fade. The production based model identifies the transformation applied to the shots as a result of these edit operations. The transformations are either in the pixel space or the color space of the video frames. Different techniques have been developed for parsing compressed video [96, 133]. [96] identifies video shots in motion JPEG based on correlation between the DCT coefficients of video frames. Algorithms for parsing MPEG coded video are introduced in [133]. It also discusses identification of video camera operations.

4

MULTIMEDIA DATA ACCESS

Access to multimedia information must be quick so that retrieval time is minimal. Data access is based on metadata generated for different media composing a database. Metadata must be stored using appropriate index structures to provide efficient access. Index structures to be used depend on the media, the metadata, and the type of queries that are to be supported as part of a database application. In this chapter, we discuss the types of indexing mechanisms that can be employed for multimedia data access.

4.1 ACCESS TO TEXT DATA

Text metadata consists of index features that occur in a document as well as descriptions about the document. For providing fast text access, appropriate access structures have to be used for storing the metadata. Also, the choice of index features for text access should be such that it helps in selecting the appropriate document for a user query. In this section, we discuss the factors influencing the choice of the index features for text data and the methodologies for storing them.

Selection of Index Features : The choice of index features should be in such a way that they describe the documents in a possibly unique manner. The definitions *document frequency* and *inverse document frequency* describe the characteristics of index features. The document frequency $df(\phi_i)$ of an indexing feature ϕ_i is defined by the number of documents in which the indexing feature appears, where $df(\phi_i) = |\{d_j \in D | ff(\phi_i, d_j) > 0\}|$. Here, d_j refers to the jth document where the document index occurs, D is the set of all documents and

$ff(\phi_i, d_j)$ is the feature frequency. This feature frequency denotes the number of occurrences of the indexing feature ϕ_i in a document d_j. On the other hand, the inverse document frequency ($idf(\phi_i)$) of an indexing feature ϕ_i describes its specificity. The inverse document feature is defined by : $idf(\phi_i) = log(\frac{n+1}{df(\phi_i)+1})$, where n denotes the number of documents in a collection. The selection of an indexing feature should be such that $df(\phi_i)$ is below an upper bound, so that the feature appears in less number of documents thereby making the retrieval process easier. This implies that the inverse document frequency $idf(\phi_i)$ for the selected index feature ϕ_i will be high.

Methodologies for Text Access : Once the indexing features for a set of text documents are determined, appropriate techniques must be designed for storing and searching the index features. The efficiency of these techniques directly influence the response time of search. Here, we discuss the following techniques :

- **Full Text Scanning** : The easiest approach is to search the entire set of documents for the queried index feature(s). This method, called *full text scanning*, has the advantage that the index features do not have to be identified and stored separately. The obvious disadvantage is the need to scan the whole document(s) for every query.

- **Inverted Files** : Another approach is to store the index features separately and check the stored features for every query. A popular technique, termed *inverted files*, is used for this purpose.

- **Document Clustering :** Documents can be grouped into clusters, with the documents in each cluster having common indexing features.

4.1.1 Full Text Scanning

In full text scanning, as the name implies, the query feature is searched in the *entire* set of documents. For boolean queries (where occurrences of multiple features are to be tested), it might involve multiple searches for different features. A simple algorithm for feature searching in a full text is to compare the characters in the search feature with those occurring in the document. In the case of a mismatch, the position of search in the document is shifted right once, and this way the search is continued till either the feature is found in the document or the end of document is reached. Though the algorithm is very simple, it suffers from the number of comparisons that are to be made

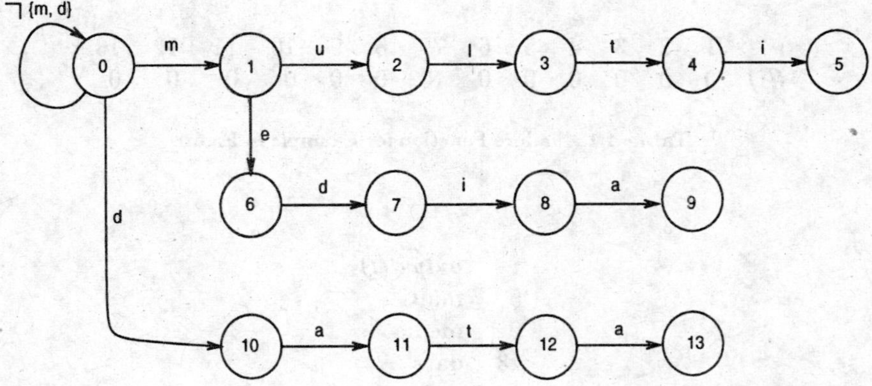

Figure 4.1 FSM for String Matching

for locating the feature. If *m* is the length of the search feature and *n* is the length of the document (in bytes), then $O(m * n)$ comparisons are needed in the worst case. Some variations of this algorithm can be used to improve the speed of search. These variations basically try to identify how efficiently one can move the position of the text pointer in the case of a mismatch. One way is to predict the location of mismatch and move the text pointer appropriately. Another approach is to do the string comparison from right to left, and in the case of a mismatch shift the text pointer right by *m* positions.

Full Text Scanning and Retrieval Using Finite State Machine : A Finite State Machine (FSM) can be used for matching the index feature (a string of characters) with the text document(s). The construction of the FSM for string matching involves the following steps.

1. Defining *Goto* function. This function defines the transition of the FSM, on receiving an input symbol, to another state. The *Goto* function reports *fail* when the transition from a state for an input symbol is undefined.
2. Defining a *Failure* function. This function is consulted when the Goto function reports *fail.* The failure function defines the transition from a state to another state, on receipt of the *fail* message. After this failure transition, the Goto function for the new state with the same input symbol is executed.
3. Defining an *Output* function. The FSM has a set of output states and the output function defines the keyword identified by each output state.

i	1	2	3	4	5	6	7	8	9	10	11	12	13
$f(i)$	0	0	0	0	0	0	10	0	0	0	0	0	0

Table 4.1 Failure Function for Example in Figure

i	*output(i)*
5	multi
9	media
13	data

Table 4.2 Output Function for Example in Figure

Consider text access with index features defined by a set {*multi, media, data*}. The Goto function for identifying these keywords is shown in Figure 4.1. The failure function can be defined as shown in Table 4.1. The failure function in this example is simple with all the states (except 7) being mapped to the initial state. For state 7, the fail state is mapped to state 10, since the character *d* has been received by the state 10 also. The output function for this FSM can be defined as shown in Table 4.2.

Full text scanning approach has the advantage that no separate search information (such as index files) has to be maintained for the documents. However, the number of comparisons to be made for searching the entire set of documents can limit the performance of the retrieval operation badly.

4.1.2 Inverted Files

Inverted files are used to store search information about a document or a set of documents. The search information includes the index feature and a set of *postings*. These postings point to the set of documents where the index features occur. Figure 4.2 shows a typical structure of an inverted file. Access to an inverted file is based on a single key and hence efficient access to the index features should be supported. The index features can be sorted alphabetically or stored in the form of a hash table or using sophisticated mechanism such as B-trees.

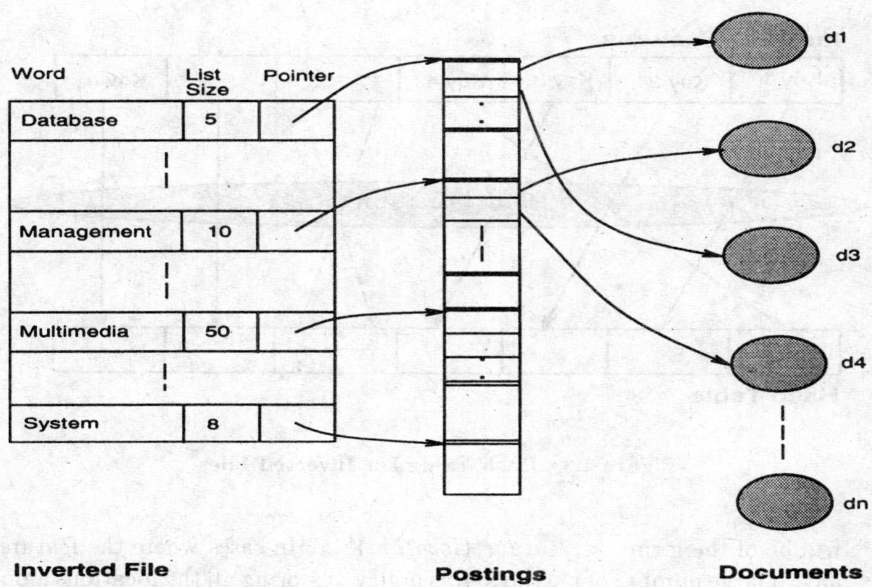

Figure 4.2 Inverted File Structure

B-trees : B-tree is an n-ary branched balanced tree. The easiest approach to construct inverted index files using B-trees is to store the tuple, $< feature, location >$, as a single entry. The disadvantage with this approach is that the tree will have multiple entries for multiple occurrences of the same feature. The following issues have to be kept in mind while constructing inverted index files using B-trees.

- Time required to access the posting for a given feature
- The ease of incrementally updating the index file
- Amount of storage required for the index file

Following approaches are used to improve the inverted index files representation.

1. Store the list of locations of occurrences of the feature instead of storing just one location with the feature. This approach removes the redundant requirement of multiple occurrences of the feature. Hence, the stored tuple

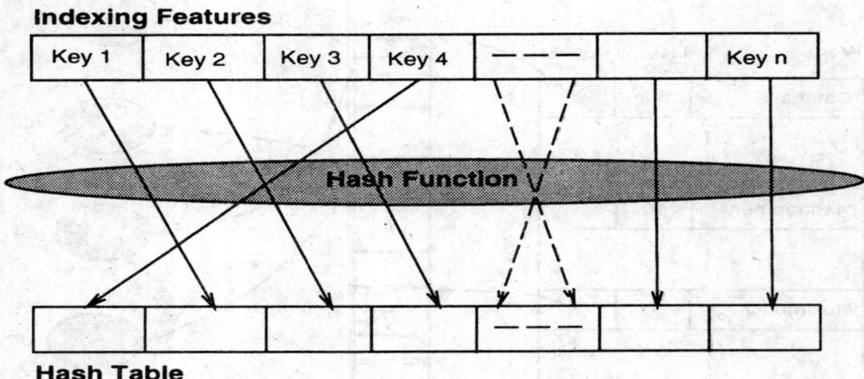

Figure 4.3 Hash Table For Inverted Files

will be of the form : $< feature, (location)^* >$. In cases where the features have a large number of postings, this policy of storing all the locations along with the feature might cause problems in terms of storage space required. An alternate approach will be to store the tuple, $< feature, pos >$, where *pos* is a pointer to a *heap file* that stores the locations of all the occurrences.

2. Using separate heap files to store the locations of all the occurrences of a feature necessitates another disk access to read the heap file. A *pulsing* technique can be used to reduce this overhead. In this technique, use of a heap file for storing the locations of occurrences is resorted to, only when the number of locations exceed a threshold t.

3. A technique, called *delta encoding*, can be used to reduce the spatial requirement for storing the locations of occurrences. Here, instead of storing the absolute values of the locations, the differences between them are stored.

4. For dynamic updates, a *merge update* technique can be adopted, where the postings are maintained in the buffer and are merged with the B-tree when the buffer becomes full.

Hash Tables : Inverted indices can also be stored in the form of a hash table. Here, a hashing function is used to map the index features that are in the form of characters or strings, into hash table locations. Figure 4.3 shows the use of hash tables for storing the feature index identifiers and the corresponding postings.

Multimedia	100	010	001	011
Database	010	001	100	010
Management	001	100	010	001
System	110	011	101	011
Signature	111	111	111	011

Table 4.3 Superimposed Coding for Multiattribute Retrieval

Text Retrieval Using Inverted Files

Index features in user queries are searched by comparing them with the ones stored in the inverted files, using B-tree searching, or hashing depending on the technique used in the inverted file. The advantage of inverted files is that it provides fast access to the features and it reduces the response time for user queries. The disadvantage is that the size of the inverted files can become very large when the number of documents and the index features become large. Also, the cost of maintaining the inverted files (updating and reorganizing the index files) can be very high.

4.1.3 Multiattribute Retrieval

When a query for searching a text document consists of more than one feature, different techniques must be used to search the information. Consider a query used for searching a book titled 'Multimedia database management systems'. Here, four key words (or *attribute values*) are specified : 'multimedia', 'database', 'management', and 'systems'. Each attribute is hashed to give a bit pattern of fixed length and the bit patterns for all the attributes are superimposed (boolean *OR* operation) to derive the *signature* value of the query. Table 4.3 shows the derivation of the signature value for the query features : *multimedia database management systems*. Table 4.3 assumes a signature size of 12 bits.

The signature value *111 111 111 011* is used as the search information for retrieving the required text document with index features *multimedia database management system*. Alternate techniques such as concatenation of the signature of individual index features (instead of the boolean OR operation), are also used. For information retrieval, more than one level can be used to store

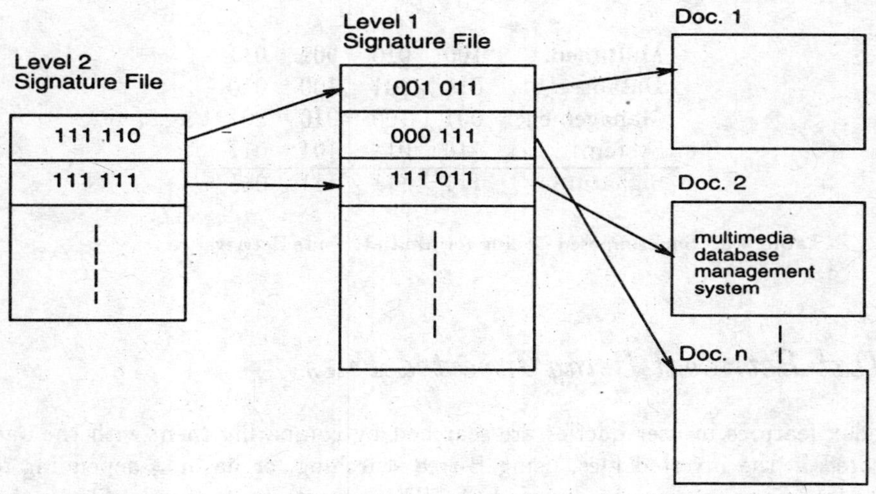

Figure 4.4 Multiple Levels of Signature Files

the signature values. Figure 4.4 shows one possibility by using two levels of signatures, with 6 bits each.

4.1.4 Clustering Text Documents

Clustering or grouping of similar documents accelerates the search since closely associated documents tend to be relevant to the same requests. The clustering principle may also be applied to the index features, instead of the documents. From the point of view of clustering, the documents, the index features and the search query are viewed as points of a m-dimensional space. The document descriptor d_j is defined as, $d_j = (a_{1,j}, ..., a_{m,j})$, where m represents the number of indexing features and $a(i,j)$ represents the weight associated with each feature. These weights must be high if the feature characterizes the document well and low if the feature is not very relevant for the document. Figure 4.5 describes the clustering of documents using weight functions. The clusters, $\{c_1, ..., c_n\}$, can be the set of index features used to characterize the document set. For example, c_1 can represent the documents where the index feature *multimedia* occurs. The weights associated with the documents (d_1 and d_3) denote the relevance of the feature *multimedia* for the two documents. If d_3's association with the feature multimedia is marginal, then the weight associated with (d_3, c_1) will be very low.

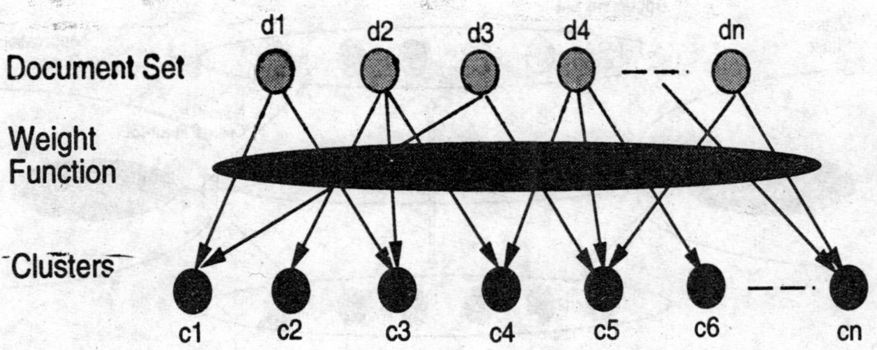

Figure 4.5 Clustering of Text Documents

The following weight functions are proposed in the literature for generating document clusters.

- Binary document descriptor : Presence of a feature by 1 and absence by 0.
- Feature frequency, $ff(\phi_j, d_j)$.
- Document frequency, $df(\phi_i)$.
- Inverse document frequency or the feature specificity, $idf(\phi_i)$.
- $ff(\phi_i, d_j) * R_j$, where R_j is the *feature relevance factor* for a document j.

The values for the above weight functions have to be estimated for generating document clusters. Weight functions based on binary document descriptor, feature frequency, document frequency and inverse document frequency are straight forward estimates of some property of index features. For example, binary document vector estimates only the presence or absence of a feature. The functions such as feature frequency, document frequency and inverse document frequency can be estimated based on the discussions in the beginning of Section 4.1. For the weight function based on the *feature relevance factor for a document*, the relevance factor has to be estimated by using one of the *learning-based approaches* discussed below.

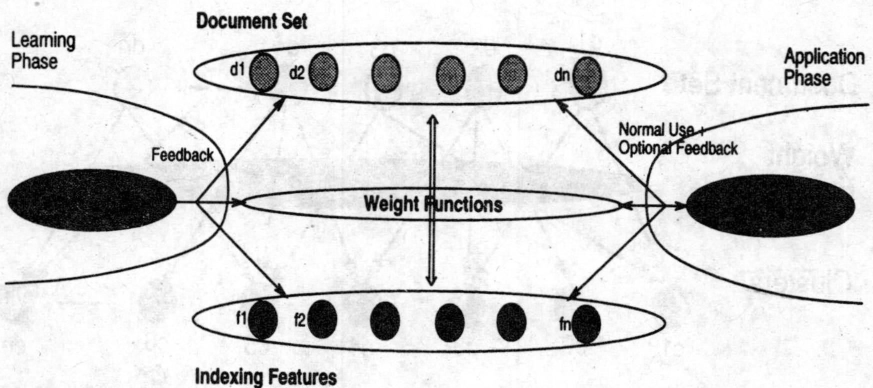

Figure 4.6 Learning Approaches for Clustering

Learning-Based Approaches For Weight Functions

Many of the learning-based methods are probabilistic in nature. Figure 4.6 describes the general principle of the learning approaches. The learning approaches have two phases : *learning phase* and *application phase*. In the learning phase, a set of *learning* queries are used to derive a feedback information. These learning queries are similar to the ones used normally for text access and they can be applied to a specific document or a set of documents. Based on the relevance of these queries for selecting document(s), probabilistic weights are assigned to the indexing features or to the documents (or both). During the application phase, *normal* queries are answered based on the weights estimated during the learning phase. The feedback information can also be derived from the *normal* queries for modifying the associated weights (as indicated by the double headed arrows for normal queries in Figure 4.6). The following methods are normally used for deriving the feedback information.

Binary Independence Indexing : In this approach, the probabilities for indexing features are estimated during a *learning phase*. In this learning phase, sample queries for a specific document d_j are analyzed. Based on the indexing features present in the sample queries, the probabilistic weights for each feature is determined. The disadvantage of this approach is that the feedback information derived from the sample set of queries is used for processing *all* the queries that occur. Since the sample set of queries cannot reflect the nature of all possible queries, the weights derived using this type of feedback may not be accurate.

Darmstadt Indexing Approach : The difference in this approach is that the feedback information is derived during the learning phase as well as the application phase. Hence, new documents and new index features can be introduced into the system. The system derives the feedback information continuously and applies it to the newly introduced components (documents or index features). Since the size of the learning sample continually increases over the period of operation, the estimates of the weight functions can be improved.

Text Retrieval From Document Clusters : Text retrieval from document clusters employ a *retrieval function* which computes the similarity measure of the index features with those described for the stored documents. The retrieval function depends on the weight functions used to create the document clusters. Documents are ranked based on the similarity of the query and the documents, and then they are presented to the user.

4.1.5 Summary

Text access is performed by queries which operate on the metadata. The text metadata, comprising of the index features and the document descriptions, has to be stored using appropriate access structures so as to provide efficient document access. We discussed approaches that use Finite State Machines (FSM) for text data access. The FSM approach does not require the index features to be stored separately. However, the entire document has to be scanned for every query using the FSM technique. Other approaches discussed include inverted files and hash tables for storing the index features and the corresponding list of documents. Cluster generation methodologies are also used to group *similar* documents. The similarity among documents are determined using weight mapping functions. We also described the techniques that are used for the weight mapping functions. Table 4.4 summarizes the techniques used for text data indexing.

4.2 ACCESS TO SPEECH DATA

The indexing features used for access to speech documents have to be derived using the methodologies discussed in Section 3.3. In terms of storage and access structures for the index features, the techniques used for text can be applied with some minor modifications. There are however some additional constraints on the choice of the index features that can be used.

Text Access Method	Technique Description
Full Text Scanning	Use FSM approach
Stored Index Features	Inverted Files : B-trees, Hash tables based
Cluster generation	Grouping similar documents using weight mapping functions

Table 4.4 Text Indexing

- The number of index features have to be quite small, since the pattern matching algorithms (such as HMM, neural networks model and dynamic time warping) used to recognize the index features are expensive. The reason is that large space is needed for storing different possible reference templates (required by the pattern matching algorithms), for each index feature.

- The computation time for training the pattern matching algorithms for the stored templates is high. For a feature to be used as an index, its document frequency $df(\phi_i)$ should be below an upper bound, as discussed in Section 4.1. However, for speech data, the $df(\phi_i)$ should be above a lower bound, so as to have sufficient training samples for the index feature.

From the point of view of the pattern matching algorithms and the associated cost, words and phrases are too large a unit to be used as index features for speech. Hence, subword units can be used as speech index features. Choice of subword units for speech index features are discussed in [127]. The following steps help in identifying and using the index features:

- Determine the possible subword units that can be used as speech index feature

- Based on the document frequency values $df(\phi_i)$, select a reasonable number (say, around 1000) index features

- Extract different pronunciations of each index feature from the speech document

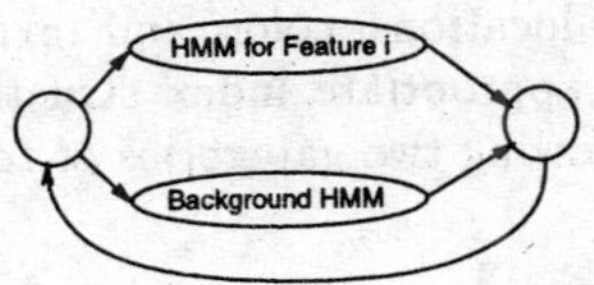

Figure 4.7 HMM for Speech Indexing

- Using the different pronunciations, train the pattern matching algorithm for identifying the index features

4.2.1 Retrieval of Speech Documents

Retrieval of speech documents is done by matching the index features given for searching and the ones available in the database. Pattern matching algorithms discussed in Section 3.4 are used for this purpose. For instance, if we are to use HMMs as the pattern matching algorithm, then each index feature selected using the above criteria are modeled by a HMM (as discussed in Section 3.4). The HMMs of all the selected index features are grouped to form a background model. This model represents all the subword units that occur as part of the speech data. Retrieval is done by checking whether a given word or sentence appears in the available set of documents. The given word or sentence for searching is broken into subword units. These units are again modeled by HMMs. The HMMs for the given index features and the background model are concatenated in parallel, as shown in Figure 4.7. The speech recognition algorithm discussed in Section 3.3.1 checks whether the HMM for the index feature occurs in the background model of the speech data. In a similar manner, other pattern matching algorithms (Dynamic Time Warping and Artificial Neural Networks) can be used for retrieving speech documents.

One can use techniques such as inverted files or signature files to store the selected index features. The retrieval strategies adopted for text can be used for speech as well.

4.3 ACCESS TO IMAGE DATA

In the previous chapter, we described the methodologies for generating metadata associated with images. Image metadata describes different features such as identified objects, their locations, color, and texture. The generated metadata has to be stored in appropriate index structures for providing ease of access. In general, the following two categories of techniques are used to store image metadata.

- Logical structures for storing the locations and the spatial relationships among the objects in an image.
- Similarity cluster generation techniques where images with similar features (such as color and texture) are grouped together such that images in a group are more similar, compared to images in a different group.

4.3.1 Image Logical Structures

Different logical structures are used to store the identified objects in an image and their spatial relationships. After the preprocessing of an image (using techniques discussed in Section 3.4.1), objects in the image are recognized and their symbolic names are assigned. Storing the identified objects involves identification of their geometrical boundaries as well as the spatial relationships among the objects.

Identifying Geometric Boundaries

Geometric boundary of objects can be stored using *Minimum Bounding Rectangle* (MBR) or by using a *plane sweep* technique which generates polygonal approximations of the identified objects.

MBR Representation : MBR is a representation that describes an object's spatial location using the minimum sized rectangle that completely bounds an object. The MBR concept is very useful in dealing with objects that are arbitrarily complex in terms of their boundary shapes. This representation can also be useful in identifying the overlaps of different objects, by comparing the coordinates of the respective MBRs. Figure 4.8(a) shows the MBR approximation for a facial image.

Sweep Line Representation : *Sweep line representation* is a technique used for identifying the geometric boundary of the objects. Here, a *plane sweep* technique is used where a horizontal line and a vertical line sweep the image from top to bottom (horizontal sweep) and from left to right (vertical sweep). A set of pre-determined points in the image called *event points* are selected so as

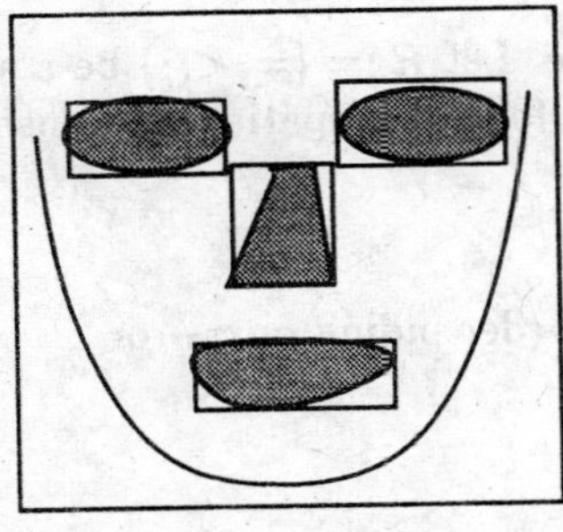

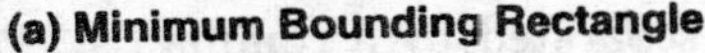

(a) Minimum Bounding Rectangle (b) Single Sweep Line

Figure 4.8 MBR and Sweep Line Representation of Objects' Spatial Organization

to capture the spatial extent of the objects in the image. The horizontal and the vertical sweep lines stop at these event points, and the objects intersected by the sweep line are recorded. Figure 4.8(b) shows the sweep line representation of a facial image. Here, the facial features such as eyes, nose, and mouth are represented by their polygonal approximations. The vertices of these polygons constitute the set of event points. If we consider the horizontal sweep line (top to bottom), the objects identified are : eyes, nose and mouth. Similarly, for the vertical sweep line (left to right), the identified objects are : left eye, mouth, nose and right eye.

Identifying the Spatial Relationships

Various techniques are used to identify the spatial relationships among the objects in an image. Here, we discuss the following techniques :

- 2D-Strings
- 2D-C Strings

2D-Strings : 2D-strings is used to represent the spatial relationships among objects in an image by representing the projection of the objects along the x and y axes. The objects are assumed to be enclosed by a MBR with their boundaries parallel to the horizontal ($x-$) and the vertical ($y-$) axis. The reference points of the segmented objects are the projection of the objects' centroids on the $x-$ and the $y-$ axis. Let $S := \{O_1, O_2, ..., O_n\}$ be a set of symbols of the objects that appear in an image. Let $R := \{=, <, :\}$ be a set of relation operators. These operators specify the following spatial relationships :

$=$	At the same spatial location
$<$	To the west of *or* to the south of (depending on $x-$ or $y-$ axis).
$:$	In the same set as

A 2D-string is represented as two substrings separated by a comma (,). The first substring describes the spatial relationships along the x axis and the second substring the relationships along the y axis. Consider the facial image shown in Figure 4.8. Let $S = \{LE, RE, N, M\}$, where LE represents the left eye, RE the right eye, N the nose, and M the mouth. The 2D string for this spatial image is : $\{LE < N : M < RE, M < N < LE : RE\}$. (The 2D-string representation of almost all facial images will be the same ! This example is used only to illustrate the use of 2D-strings.) Thus, a 2D-string can be thought of as the symbolic projection of the identified objects in an image along the $x-$ and the $y-$axis. The disadvantage of 2D strings is that the spatial relationships among the objects are represented based on the projection of the objects' centroids onto the $x-$ and $y-$axis. This projection of objects' centroids alone do not reflect the complete picture of the spatial organization.

2D-C Strings : This approach overcomes the disadvantage of 2D-strings by representing spatial relationships among the boundary of objects (instead of objects' centroids as in 2D-strings). There are thirteen possible relationships between two rectangles that enclose the objects (ignoring the rectangles' length information) along the $x-$(or $y-$)-axis. 2D-C strings describes these spatial relationships by using a set of operators $\{<, =, |, \%, [,], /\}$, as shown in Figure 4.9.

Retrieval Based Spatial Relationships

The retrieval of images based on their spatial relationships depends on the technique used to represent the relationships. Consider the case of a query-by-example. In the case of sweep line representation of the spatial relationships, the sweep line representation is generated for the query image also. If the generated representation for the query image matches the one(s) stored in the database, then the image(s) is(are) retrieved.

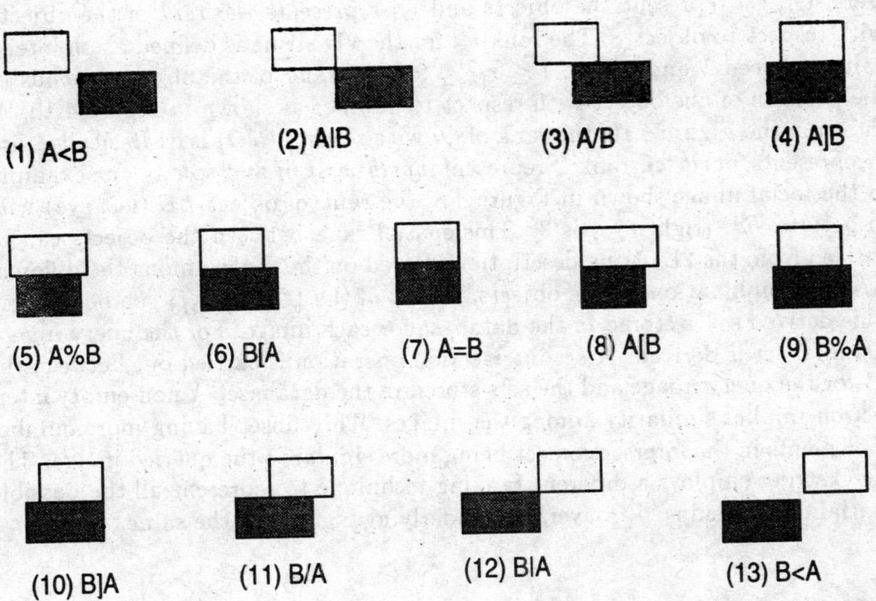

Figure 4.9 13 Types of Spatial Relations in One Dimension

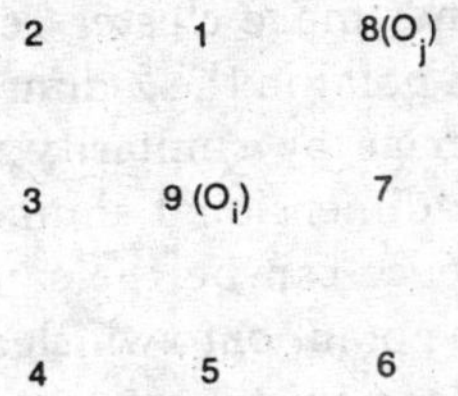

2	1	8(O_j)
3	9 (O_i)	7
4	5	6

Figure 4.10 Spatial Rank of Objects

For techniques such as 2D- or 2D-C string, pre-processing is required to *translate* the string description of each image into a set of the form : $\{O_i, O_j, r_{ij}\}$. Here, O_i, O_j represent the objects and r_{ij} represents the *rank* of the object i with respect to object j. The rank r_{ij} for the 2D-string is defined as an integer value between 1 and 9, i.e., $1 \leq r_{ij} \leq 9$. The rank of the object depends on the position of one object with respect to another as shown in Figure 4.10. As shown in the Figure 4.10, the rank of O_i with respect to O_j is 8. Basically, rank 1 represents *north of*, rank 2 represent *north-west of* and so on. For example, in the facial image shown in Figure 4.8, the rank of object LE (left eye) with respect to RE (right eye) is 3. This spatial rank between the objects can be derived from the 2D-string description. Based on the ranks among the different possible combinations of the objects, the set of the $\{O_i, O_j, r_{ij}\}$ can be derived. This derived set is stored in the database for each image. For the query image, a similar set is derived. A set intersection operation is carried out between the set for the query image and the sets stored in the database. A non-empty intersection implies similarity among the images. The subset having more number of elements in the intersection set being more similar to the queried image. The 2D-C string employs a different ranking technique to represent all the possible spatial relationships. However, the underlying concept is the same.

4.3.2 Image Cluster Generation Methodologies

Features such as color and texture of an image can be indexed using similarity cluster generation methodologies. Here, a mapping function is defined to generate a similarity measure based on the features to be indexed. Images are then grouped in such a way that the difference between the similarity measures of the images within a cluster are below a known upper bound. Figure 4.11 describes the basic idea behind image cluster generation. Here, the mapping function F maps an image to a point in the 2-dimension *similarity space*. Hence, a query trying to retrieve image by similarity within a distance d becomes a circle of radius d in the 2-dimensional similarity space. In practice, the dimension of the similarity space can be the same as the number of features used, calling it a *f-d space*. The point onto which an image is mapped in the f-d space is called.a *f-d point*. Image cluster generation methodologies necessitate the following steps.

- Definition of a mapping function F for the features based on which images are to be indexed

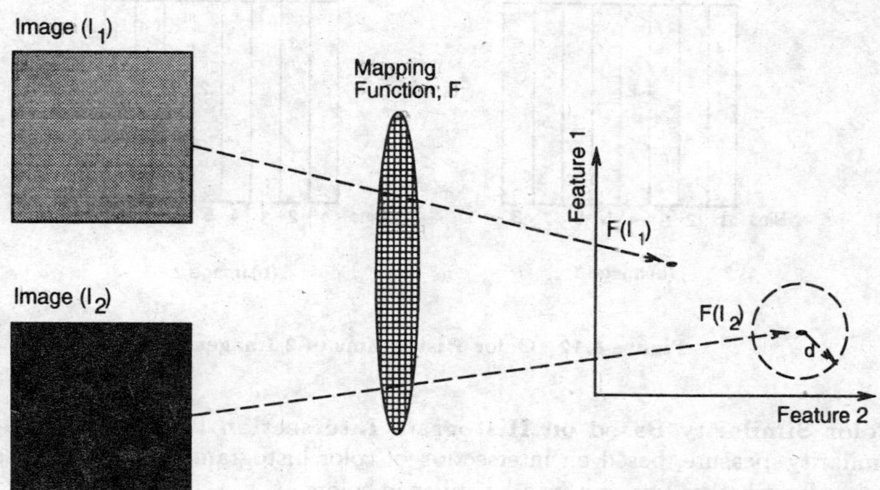

Figure 4.11 Basic Idea Behind Image Clustering

- Use of spatial access structure to group the f-d points and to store them as clusters

The mapping function should be able to map an image to a *f-d* point in the similarity space. It should also be able to *preserve* the *distance* between two images, i.e., if the dissimilarity between two images can be expressed as quantity D, then the mapping function should map the two images onto the similarity space such that the two points are a distance δ apart, where δ is proportional to D. Preserving the distance in the similarity space makes sure that two dissimilar images cannot be misinterpreted as similar. Mapping functions depend on the feature to be indexed. Now, we discuss mapping functions for image features such as color and texture.

Color Mapping Functions

Similarity between two images can be estimated based on extracted color features as well as the spatial locations of the color components in the images. Here, the spatial information implies the positions of pixels having the same color. Most of the color mapping functions work on the extracted color features and do not consider the spatial information. The extracted color features are stored in the form of histograms, as discussed in Section 3.5.1.

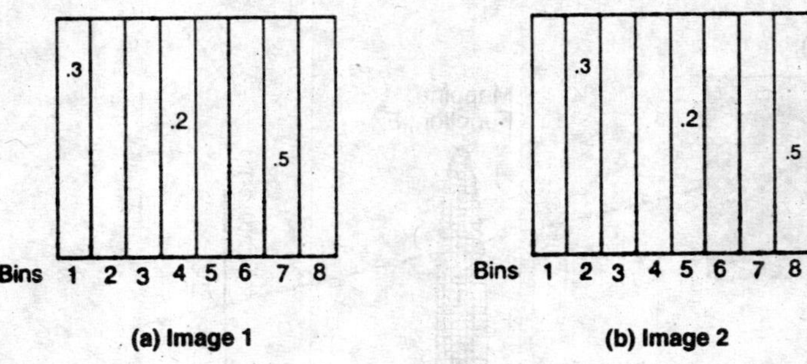

Figure 4.12 Color Histograms of 2 Images

Color Similarity Based on Histogram Intersection : The following similarity measure, based on intersection of color histograms, can be used to determine whether two images are similar in colors.

$$Sim(I_1, I_2) = \frac{\Sigma_{i=1}^{b} min(I_{1_i}, I_{2_i})}{\Sigma_{i=1}^{b} I_{2_i}}$$

Here, $Sim(I_1, I_2)$ is the *distance* between the two images I_1 and I_2 in the similarity space, I_{1_i} and I_{2_i} are the number (or the percentage) of pixels in the ith color bin of images of I_1 and I_2 respectively. Here, b denotes the number of color bins describing the color shades that are distinguished by the histogram. Based on the mapping function, two exactly similar images have a similarity measure of 1. For instance, consider the color histograms of two images shown in Figure 4.12. The two images have pixels in adjacent color bins but not in the same bins. Hence, $min(I_{1_i}, I_{2_i})$ is zero for all the color bins i. Hence, the above similarity measure gives a value of zero for the two example color histograms. The disadvantage with this similarity measure is that, only the number of pixels in the *same* color bin are compared. It does not consider the correlation among the color bins. In the above example, if we assume that adjacent color bins represent shades of a similar color, then the two images might be more similar looking. Hence, it will not be fair to give the two images a similarity measure of zero.

Color Similarity Using Correlation : In order to take into consideration similarity among different color shades, a similarity measure taking into account the correlation among the colors can be used. This similarity measure is defined by the following function :

$$Sim(I_1, I_2) = \Sigma_{i=1}^{b} \Sigma_{j=1}^{b} a_{ij} (I_{1_i} - I_{2_j})(I_{1_j} - I_{2_i})$$

Here, a_{ij} is the correlation function defining the similarity between the ith and j^{th} colors. Other terms are the same as defined in the previous mapping function. As an example, for the color histograms shown in Figure 4.12, let us assume that a_{ij} is 0.5 for adjacent color bins and 0 for other bins. Then, the similarity measure for the two images in Figure 4.12, will be 0.190.

Texture Mapping Functions

Texture features such as coarseness, contrast and directionality are also used to characterize images. Coarseness is described by terms such as fine, coarse, etc. Coarseness measure is defined by considering the variations in the gray-levels and elements size. Contrast is the description of gray level distribution in an image. Directionality describes the orientation of the patterns that appear in an image. (The similarity measures that can be used for texture features such as coarseness, contrast, and directionality are described in [6]. Cluster generation functions for image textures can be defined in the three dimensional texture space corresponding to coarseness, contrast, and directionality. Interested readers can refer [145] for further details).

Color and Texture Indexing

The techniques described above basically help in mapping the features of images onto points in a similarity space. These points have to be stored using appropriate access structures so that their fast retrieval can help in fast query processing. Typically, range trees, called *R-trees* are used to store this multi-dimensional space point information.

R-trees : R-tree is a height-balanced tree, an extension of B-trees for multidimensional objects. Here, a node in the R-tree can be assumed to represent a minimum bounding rectangle (MBR). MBR represented by a parent node contains the MBRs represented by its children. Leaf nodes in the R-trees have pointers to the objects that fall within the MBR represented by the individual nodes. R-trees can be represented by the tuple (N_t, T, E, b_f), corresponding to the following.

- N_t represents the non-leaf nodes. These nodes contain entries of the form (l, ptr), where l is the MBR that covers all rectangles in a child node and ptr is a pointer to a child node in the R-tree.

- T represents the leaf nodes. These nodes contain entries of the form (l, obj_{id}), where l is the MBR covering the enclosed spatial objects and obj_{id} is the pointer to the object description.

- E represents the set of edges in the tree.

- b_f represents the branching factor of the tree.

The multidimensional feature points generated using the color or texture of images can be indexed using R-trees. The points in the similarity space are enclosed within a MBR. This partitioning can be done by setting a limit on the number of points in each base rectangle. Figure 4.13 shows the MBRs of the feature points in a 2-d space and the corresponding R-tree. Here, partitioning of the feature points are done by limiting them to four numbers within a MBR. $A1, ..., A4, B1, B2, C1, C2$ and $C3$ are the MBRs enclosing the feature points. A, B and C are the parent nodes whose MBRs enclose those represented by the leaf nodes.

Retrieval Using R-tree Feature Index : R-tree feature index, for example a color index, can be used for searching in the following cases :

- When image color is specified by its RGB (Red, Green, Blue) values. Here, the R-tree has to be accessed to find the MBR which encloses the point defined by the given RGB values. The points enclosed by the chosen MBR correspond to the images with similar color values.

- When an example image is provided. Here, the query image has to be mapped onto the similarity space first. Then, the R-tree has to be accessed to determine all the base rectangles that intersect the query rectangle. The points enclosed by the intersecting MBRs correspond to those images with similar color as the query image.

The above approach proceeds in a two-phase manner.

- In the first phase, a *quick-and-dirty* test is performed to determine a list of images that are close to the query. This quick-and-dirty test is done by selecting the MBR enclosing the possible images.

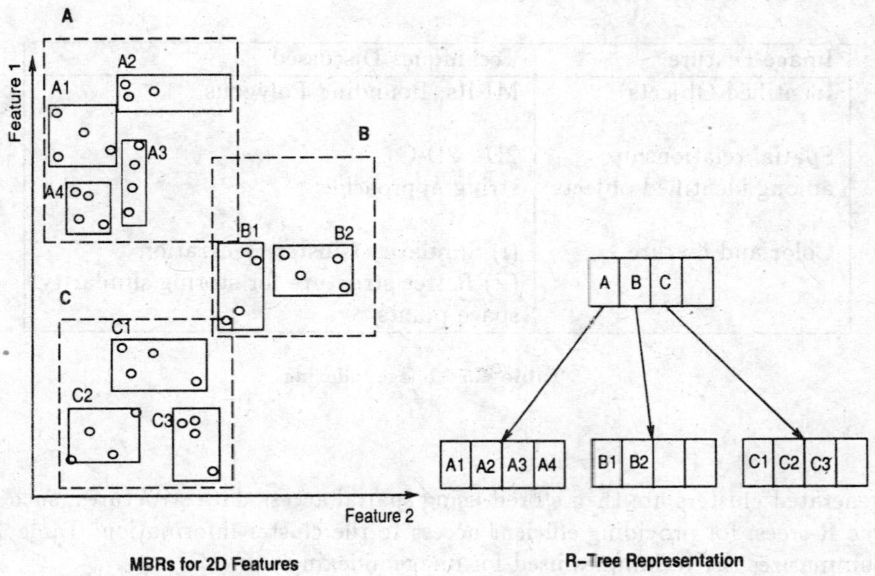

Figure 4.13 MBRs in a 2-d Similarity Space and the R- trees

- In the second phase, the images within the chosen MBR are ranked according to their similarities with the query image.

A similar technique can be to applied for processing queries to retrieve images with the same texture. The textural properties of the query image can be mapped to a *f-d* point in the similarity space. Then, the MBR enclosing the point is identified. The points inside the MBR correspond to the images with similar texture as the query image.

4.3.3 Summary

Image data access is performed by using the generated metadata : identified objects in the image, their spatial relationships, and features, such as color and texture. These metadata have to be stored in appropriate structures for providing efficient access. The geometric boundary of the identified objects in the image are stored using MBR or polygons enclosing the objects. The spatial relationships are stored using 2D- or 2D-C string approaches. Features, such as color and texture, are indexed using cluster generation methodologies. The

Image Feature	Techniques Discussed
Identified Objects	MBRs, Bounding Polygons
Spatial relationship among identified objects	2D-, 2D-C string approaches
Color and texture	(i) Similarity Cluster Generation (ii) R-tree structure for storing similarity space points

Table 4.5 Image Indexing

generated clusters are then stored using spatial access data structures, such as the R-trees, for providing efficient access to the cluster information. Table 4.5 summarizes the techniques used for image indexing.

4.4 ACCESS TO VIDEO DATA

Access to video is by means of the stored video metadata. As discussed in Section 3.5, the video metadata usually consists of identified video shots, their descriptions, the descriptions regarding the camera movements, object movements, and qualities of individual video frames, such as lighting. Appropriate access storage structures have to be used for storing these metadata in order to provide fast access. For selecting appropriate access structures to store video metadata, we need to identify the following characteristics of the metadata.

- Video shots can be described as a sequence of frames. For example, a video shot can span from frame numbers 25 to 42.
- Descriptions of video can be with respect to the objects (living and non-living) and events that occur. These objects and events can span video shots. Hence, occurrence of objects and events can also be described based on the frame sequences in which they appear.
- Other descriptions such as camera movement and object motion are more or less related to the particular video shots. Hence, they can be described based on the sequences of frames in which they appear.

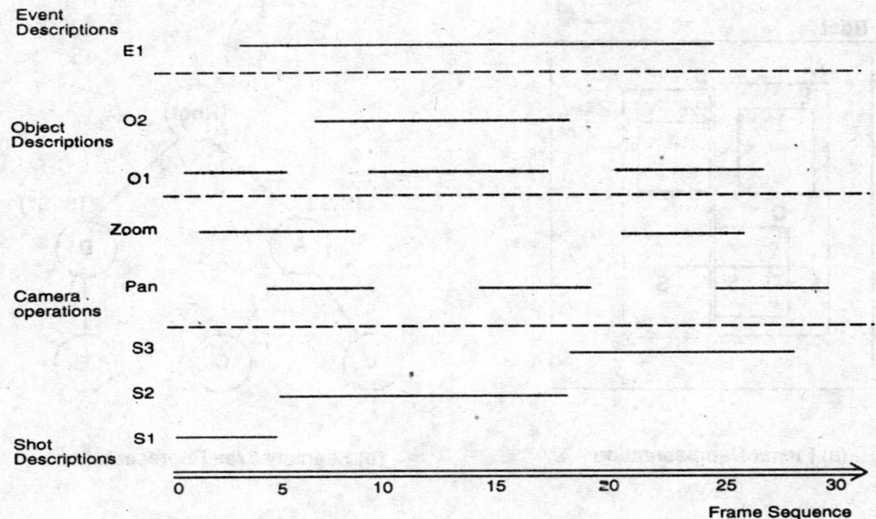

Figure 4.14 Frame Sequences and Metadata Information

Figure 4.14 shows the video metadata description over a sequence of frames. The described metadata includes camera movement, object motion, camera shots, objects in the video and event descriptions. For example, the camera operation, panning, occurs in the frame intervals [5,10], [15,20] and [25,30]. Since video metadata can be described over a sequence of frames, they can be stored in the form of an interval tree or a segment tree.

4.4.1 Segment Index Trees

Segment index trees, called the $SR-Trees$, are adaptations of the R-Tree structure for segment intervals. A node of the SR-tree stores an interval (instead of a MBR, as in the case of R-trees). Interval represented by a parent node contains the intervals represented by its children nodes. Thus, SR-trees provide efficient mechanisms to index both *interval* and *point* data in a single index (since a point is also contained by an interval). A distinct feature of the SR-Tree is that a new interval that is to be inserted into the index can be split. The split intervals can then be inserted into the tree.

Consider the planar representation of a set of segment intervals and the corresponding tree representation, shown in Figure 4.15. *SS*" is a new segment

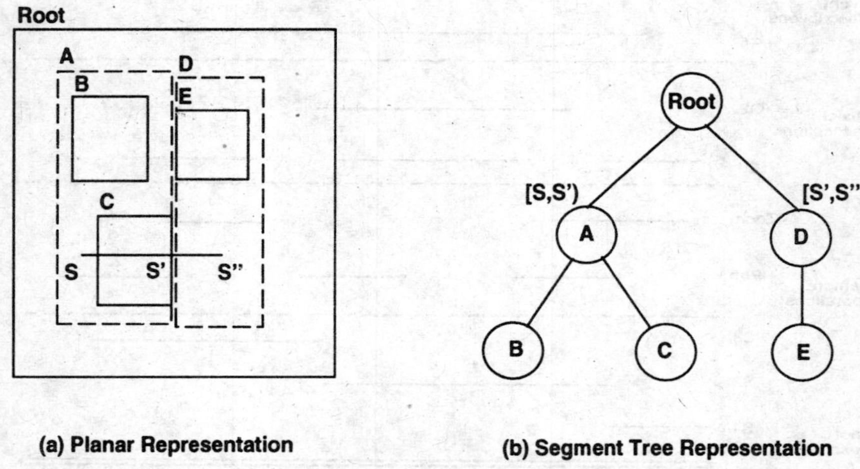

Figure 4.15 Segment Tree : Cutting a Segment

that is to be inserted into the index tree. As part of the insertion algorithm, each node N (beginning with the root-node, searched in top-down, depth-first mode) is tested to find out if the region spanned by N encompasses the new segment SS". If it does, SS" is inserted into N. In the example, SS" spans node C, but not its (C's) parent node A. Hence, SS" is cut into a *spanning portion, SS'*, (which spans node C and is fully enclosed by C's parent) and a *remnant portion, S'S"* (which extends beyond the boundary of C's parent). Then, the spanning portion (SS') is stored in node A and the remnant portion ($S'S$") is stored in node D.

Frame Segment Tree

Figure 4.16(a) shows the segment tree for storing the frame sequences that are shown in Figure 4.14. This approach, called the *frame segment tree*, is used for storing the sequences of video frames. Each node in the frame segment tree represents a *frame sequence* [x, y), starting from frame x including all frames *upto* y, but not including frame y. The list of metadata (objects, camera movements, etc.) described by the frame segment is indicated by the side of each node in the frame segment tree.

Storing Objects, Events, Camera Movements, .. : The frame segment tree described above contains all the video metadata. However, data access might be made through queries that describe the objects, the events, or the

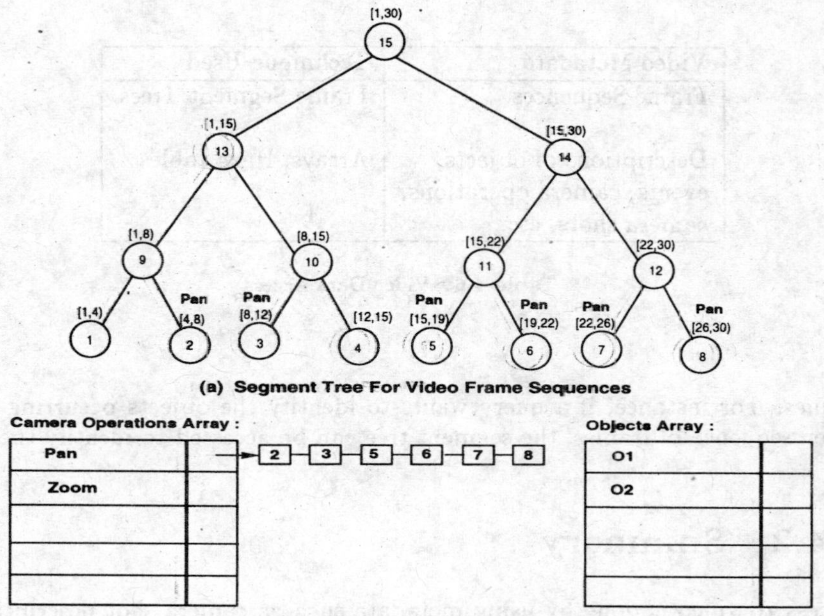

Figure 4.16 **Array-Linked Segment Tree for Video Frame Sequences**

camera operations. Hence, faster access can be provided by storing information for object and event descriptions in separate arrays. We can also use hash tables in case the number of entries in the arrays are large. These arrays store the identifiers of the metadata (objects, events, camera operations, camera shots, etc.) as well as *ordered linked list* of pointers to nodes in the segment trees, as shown in Figure 4.16(b).

Retrieval of Video Data : If queries involve descriptions of objects, events, or camera operations, then the array storing the metadata identifiers needs to be accessed first. This array gives an ordered list of the nodes in the frame segment tree. These nodes in turn, gives the sequence of video frames in which the required metadata is contained. As an example, if a query wants to retrieve the sequence of video frames where the camera operation is panning, then the camera operations array, shown in Figure 4.16(b), is first accessed. This gives us the sequences of frame segment tree nodes as : 2,3,5,6,7,8. Accessing these nodes in the tree, we get the sequence of video frames : [5,10], [10,15] and [25,30]. If queries are in such a way that the frame segment tree can be accessed directly, then the tree can be searched to get the required sequence of video

Video Metadata	Technique Used
Frame Sequences	Frame Segment Trees
Descriptions of objects, events, camera operations, camera shots, etc.,	Arrays, Hash tables

Table 4.6 Video Data Access

frames. For instance, if a query wants to identify the objects occurring in a given sequence of frames, the segment tree can be accessed to identify them.

4.4.2 Summary

Access to video is done by using metadata such as camera shot descriptions, objects occurring in video frames, object movements, camera operations, etc. These metadata are described over a sequence of video frames. These sequence of frames can be considered as intervals and hence can be indexed using segment trees or interval trees. The segment trees can also be used to index the temporal characteristics associated with media objects. For example, information regarding the time intervals in which a particular action takes place in the video can be indexed using segment trees. We described the structure of $SR - Trees$ for indexing video frame sequences. We also described the use of ordinary arrays or hash tables for storing metadata identifiers. Table 4.6 summarizes the techniques used for storing the video metadata.

4.5 CONCLUDING REMARKS

Multimedia data access is done by using the metadata for different media composing a multimedia database. These metadata have to be stored using appropriate access structures for providing efficient access. The access structures that can be used depends on the media, the metadata, and the type of queries that are to be supported as part of a database application.

Access to text data is done by queries providing index features that have to be checked for their presence in the set of stored documents. The simplest

Media	Issues	Access Structures
Text	Full Text Scanning	Use FSM approach
	Store Index Features	Inverted Files, Hash tables
	Cluster generation	Grouping similar documents using weight mapping functions
Image	Identified Objects	MBRs, Bounding Polygons
	Spatial relationship among identified objects	2D- and 2D-C string approaches
	Color and texture	(i) Similarity Cluster Generation (ii) R-tree structure for storing similarity space points
Video	Frame Sequences	Frame Segment Trees
	Descriptions of objects, events, camera operations, camera shots, etc.,	Arrays, Hash tables

Table 4.7 Access Techniques for Different Media

approach is to scan the entire set of documents. An approach based on Finite State Machines (FSM) can be used for this purpose. However, the response time of document retrieval suffers when the document size and the number of documents increase. Hence, the index features have to be stored separately. Inverted files can be used for this purpose. Similar documents can be clustered based on the occurrence of common index features. We described different approaches that are used for clustering documents. The cluster generation techniques discussed for text can be used for handling speech documents as well.

Access to image data is done by queries : (i) on the objects contained in the images, (ii) on their spatial relationships, and (iii) by example, such as retrieving images with similar color or texture. We discussed the access structures such 2D- and 2D-C for storing the identified objects and their spatial relationships.

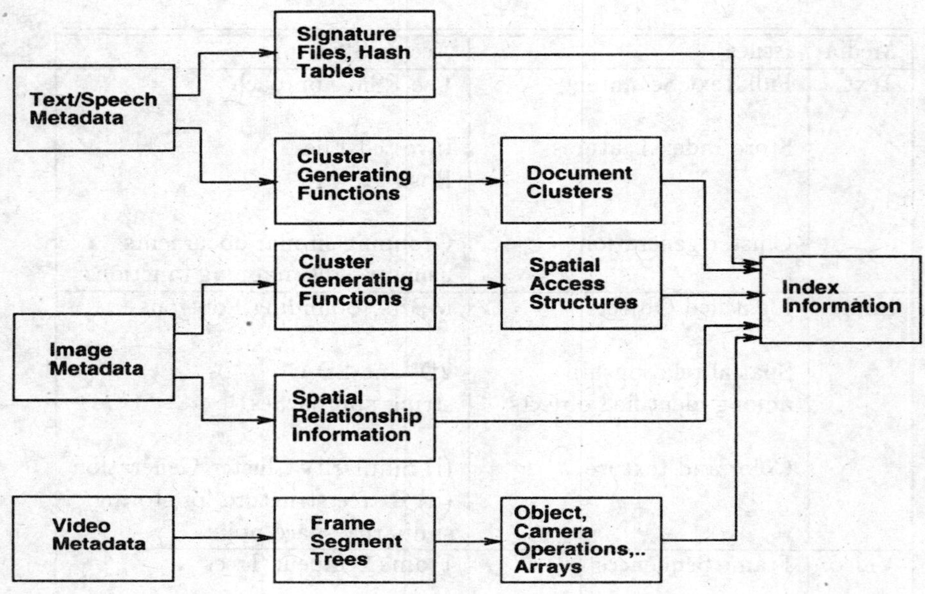

Figure 4.17 Components of Index Manager

For color and texture features, we discussed cluster generation methodologies that can be used for grouping similar images.

Video data access is done by queries on the objects or events occurring in the video, or by queries on the camera operations, camera shots, object motions, etc. These metadata are described over a sequence of video frames. These sequences can be considered as intervals. We described segment or interval tree based approaches for storing the frame sequences and the corresponding metadata. The metadata identifiers for object descriptions, event descriptions, etc., can also be stored separately using ordinary arrays or hash tables. Table 4.7 summarizes the different access structures that are used for different media.

Based on the above discussions, we can visualize a simple index manager as shown in Figure 4.17. Text and speech metadata are similar in nature. They can be indexed using mechanisms such as signature files or hash tables. Alternatively, cluster generating functions can be used to generate document clusters. Image metadata is grouped by cluster generation functions and the generated clusters are indexed using spatial access structures. Spatial relationships information of objects in an image are stored using 2D- or 2D-C strings. Video metadata is stored in the form of segment trees, and arrays of camera operations and objects occurring in the video.

Bibliographic Notes

Access methods for text are surveyed in [18]. Finite State Machine (FSM) approach for matching the index features (a string of characters) with text documents is described in [2]. Algorithms, for constructing the FSM based on the keywords and for using the FSM to search the documents in a single pass, have also been suggested in [2]. Variations of the FSM approach to improve the efficiency of searching are presented in [5, 4]. Approaches have been suggested to improve the inverted index files representation in [45]. Incremental updates of inverted files are discussed in [119]. Binary Independence Indexing for text documents is introduced in [1]. Darmstadt Indexing approach for text is presented in [53]. Order preserving hash functions for information retrieval are introduced in [55].

Retrieval of speech documents are discussed in [89, 126]. For identifying index features in speech data, a *recognition model* is used [69]. Choice of subword units for speech index features are discussed in [127].

Retrieval techniques for images are discussed in [31, 51, 59, 79, 115, 152, 134, 151]. [6] presents the texture features that can be used to characterize images. The measures that can be used for texture features such as coarseness, contrast, and directionality are described in [6]. Retrieval of images based on texture features are examined in [116, 150, 145]. 2D-strings used to represent the spatial relationships among objects in an image, are described in [25]. 2D-C Strings represents spatial relationships among the boundary of the objects [75]. Other techniques such as $\theta\Re$-String [130], spatial orientation graph [136], and Skeletons [26] are also used to describe the spatial relationships among the objects in an image. Retrieval of images based on similar shapes is discussed in [51, 152].

Similarity measure, based on intersection of color histograms, is presented in [47]. Similarity measure that takes into account the correlation among different color shades, has been introduced in [31, 138]. R-trees and algorithms for searching, insertion and deletion of entries in the R-trees are described in [16]. Variations of R-trees, called R+-trees and R^*-trees, have been proposed for improving the space utilization or for improving the efficiency of insertion into the R-trees[20, 46, 80].

Segment index trees, called the $SR - Trees$, are introduced in [61, 32]. [139] describes the use of segment trees for storing video metadata.

5

MULTIMEDIA INFORMATION MODELING

Information handled by multimedia databases is composed of media objects, derived metadata, objects' temporal and spatial characteristics. This information is continuously modified and manipulated by applications. In this chapter, we discuss the modeling techniques used to define and manipulate multimedia information. First, we describe object-oriented models for multimedia databases. Then, we present models for describing temporal characteristics associated with multimedia information. Next, we discuss representations of spatial requirements of multimedia objects. Finally, we describe multimedia authoring for representing temporal and spatial requirements of multimedia objects.

5.1 OBJECT-ORIENTED MODELING

Object-oriented approach appears to be natural for most multimedia database applications. Object-oriented paradigm is based on encapsulating code and data into a single unit, termed an *object*. The encapsulated code defines the operations that can be performed on the data. The interface between an object and the rest of the system is defined by a set of *messages*. Hence, an object has the following components associated with it :

- A set of *variables* that contains the data for the object (the value of each variable itself being an object).
- A set of *messages* generated by the system, to which the object responds. However, the messages used for communication among the objects bear no relation to the computer network messages (or packets).

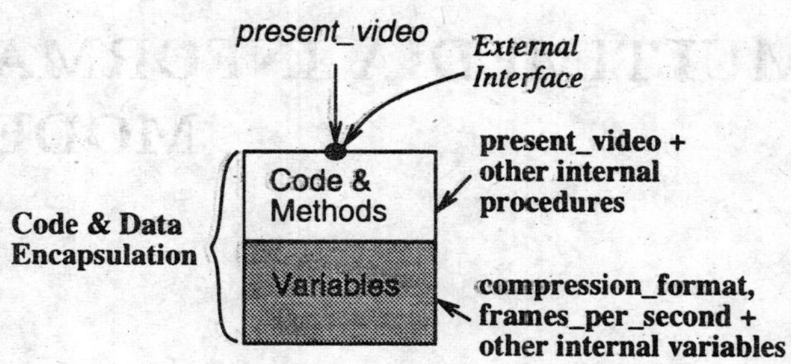

Figure 5.1 A Typical Multimedia Object

- A set of *methods* for handling messages received by an object. A method is a body of code that returns a value (or another object) as a *response* to the received message.

Figure 5.1 describes a typical multimedia object, video. The external interface is in the form of a message, *present_video*. The details of the video object are encapsulated in the form of code (*present_video* and other internal procedures) and data structures (*compression_format, frames_per_second* and other internal variables). Encapsulation of the code and data structures associated with an object helps in providing a transparent external interface to other objects. For instance, different compression methodologies can be employed for reducing the size of a video object. The object has to be uncompressed before it can be displayed on a window. The uncompression function depends on the employed compression methodology, and hence display functions for different video objects may differ. By encapsulating the code for uncompression and display within the video object, the same external interface (*present_video*) can be provided to other objects. Hence, the internal components of the video object, such as compression format, can be modified without affecting other objects. Now, we discuss the salient features of the object-oriented approach.

5.1.1 Class Hierarchy

In object-oriented modeling, similar objects are grouped to form a *class*. Similar objects have the following characteristics : they respond to the same messages, use the same methods, and have variables of same name and type. Hence, objects in a class share a common definition though the values assigned to the variables can differ. Each object is called an *instance* of its class. For representing the similarities among classes, a hierarchical description is used.

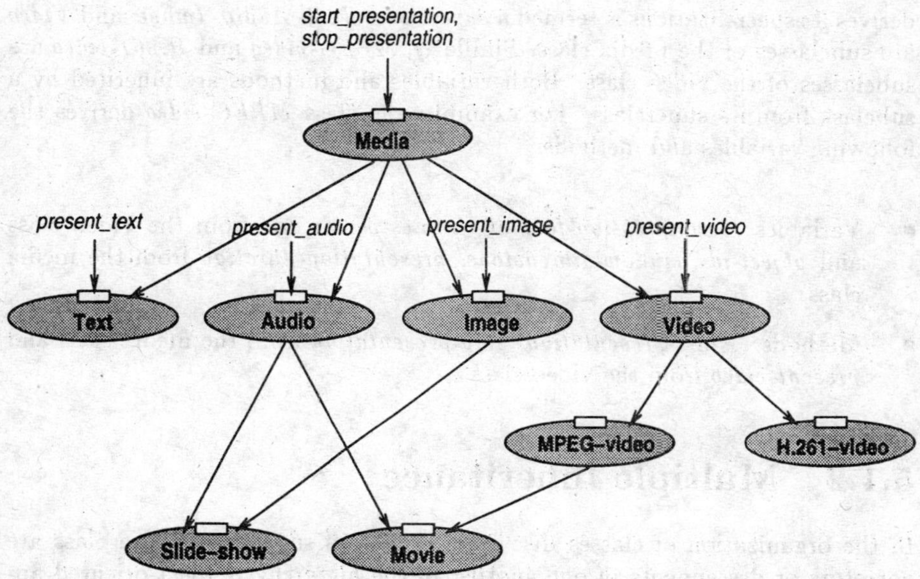

Figure 5.2 Hierarchical Structure of A Media Class

Figure 5.2 shows a possible hierarchical structure of a *media* class. In this example, the variables and methods associated with different classes can be as follows.

- **Media.** Variables : *object-id, window-dimensions, presentation-duration.* Methods : *start-presentation, stop-presentation.*

- **Text.** Variables : *text-format, font-type, page-size.* Methods : *present-text.*

- **Audio.** Variables : *compression-format, audio-type.* Methods : *present-audio.*

- **Image.** Variables : *compression-format.* Methods : *present-image.*

- **Video.** Variables : *compression-format, frames-per-second.* Methods : *present-video.*

The media class is a general representation of a media object. The specializations of a class are called the *subclasses.* The class from which a subclass derives its specializations is termed a *superclass*. *Text, Audio, Image*, and *Video* are subclasses of the media class. Similarly, *MPEG-video* and *H.261-video* are subclasses of the video class. Both variables and methods are inherited by a subclass from its superclass. For example, the class *MPEG-video* derives the following variables and methods :

- Variables : *compression-format, frames-per-second* from the video class and *object-id, window-dimensions, presentation-duration* from the media class.
- Methods : *start-presentation, stop-presentation* from the media class and *present-video* from the video class.

5.1.2 Multiple Inheritance

In the organization of classes discussed above, all superclasses of a class are ancestors or descendents of one another in the hierarchy. Object-oriented approach allows a class to inherit variables and methods from multiple superclasses. This concept, termed *multiple inheritance*, describes the class relationships by a rooted directed acyclic graph (DAG). As an example, consider a movie presentation and a slide show. Movie presentation involves presenting video and audio objects whereas a slide show involves presenting image and audio objects. Figure 5.2 shows the DAG structure reflecting the multiple inheritance of movie and slide show classes. Here, the *Movie* class inherits from the classes *MPEG-video* and *Audio*. The *Slide-show* class inherits from the classes *Image* and *Audio*.

5.1.3 Object-oriented Multimedia Data Modeling

The advantages of using object-oriented approach is that :

1. The complexity of the multimedia objects can be better modeled using object-oriented approach. The *primitive objects* of a multimedia database application are text, image, audio, and video. Image, audio, and video objects can be modeled as binary objects (*Binary Large Objects* : BLOBs). However, additional interpretations, in the form of metadata, are needed in order to interpret the information contained in them. Object-oriented

models are capable of representing the binary nature as well as the metadata associated with them.

2. Information stored as part of multimedia databases has a structure that is characteristic of each application. Object oriented approach can help in modeling the structure in a logical manner.

3. Multimedia database applications might require new data types to be defined as part of the schema. Additions and deletions of new types might also be required as part of the application. Object-oriented approach can handle this issue.

However, the current object-oriented methodologies employed for representing databases have the following drawbacks when they are applied to multimedia systems.

- Multimedia applications might need to group diverse objects and access them in a collective manner. Object-oriented programming languages normally operate on objects as individual entities. Hence, set-oriented access to objects might be needed for multimedia database applications. This can be provided by allowing the definition of a class to represent not only a type defining its characteristics that are common to its instances, but also the set of instances.

- As discussed in Chapter 3, metadata associated with multimedia objects define the interpretations associated with them. These interpretations are very much user as well as application dependent. Hence, metadata associated with multimedia objects are quite diverse in nature. Also, the entire metadata may not be described initially. Interpretations may be added, deleted, and modified dynamically. Such dynamic modifications might modify the schema associated with the database. Hence, the database schema might have to be independent of the class hierarchy.

- Edit operations in multimedia databases can involve creation of new objects that might be comprised of portions of existing objects. Existing object-oriented approaches do not allow values (such as metadata, time durations, or spatial dimensions) associated with the objects to be inherited.

5.1.4 Object Oriented Models : Case Studies

In this section, we discuss the following object-oriented models that address the above disadvantages.

- **O**bject **V**ideo **I**nformation **D**atabase.
- Jasmine Model.

OVID Model

A video database system named **OVID : O**bject **V**ideo **I**nformation **D**atabase has been described in [102]. The salient features of the OVID system are :

- The notion of a video object identifies even arbitrary sequences of frames as an independent object (a meaningful scene). For instance, frame numbers 15 to 30 and 40 to 60 can form a video object. Each object has its own attributes (metadata), and the object contents can be described in a dynamic and incremental way.
- The video database system is *schemaless*, i.e., the class hierarchy of the object-oriented database model is not assumed as a database schema. Hence, dynamic modifications of metadata or objects do not modify the database schema.
- Inheritance of the object attributes is based on an *interval inclusion* relationship. An interval is represented by a starting frame number and an ending frame number, and denotes a continuous sequence of video frames. An OVID object may be composed of multiple such intervals.

An OVID video object definition consists of : an *object identifier (oid)*, a *set of intervals (I)* and a collection of *attribute/value pairs (v)*. Figure 5.3 shows an example from the movie *Who Framed Roger Rabbit*. The clip shows the scene where the cartoon character Jessica, the rabbit, and the actor Bob Hoskins meet with an accident in the Toontown. In Figure 5.3, object o_1, with an interval I_1, describes the entire clip. The attributes associated with o_1 are as follows :

- *Event* : Jessica meeting with an accident.

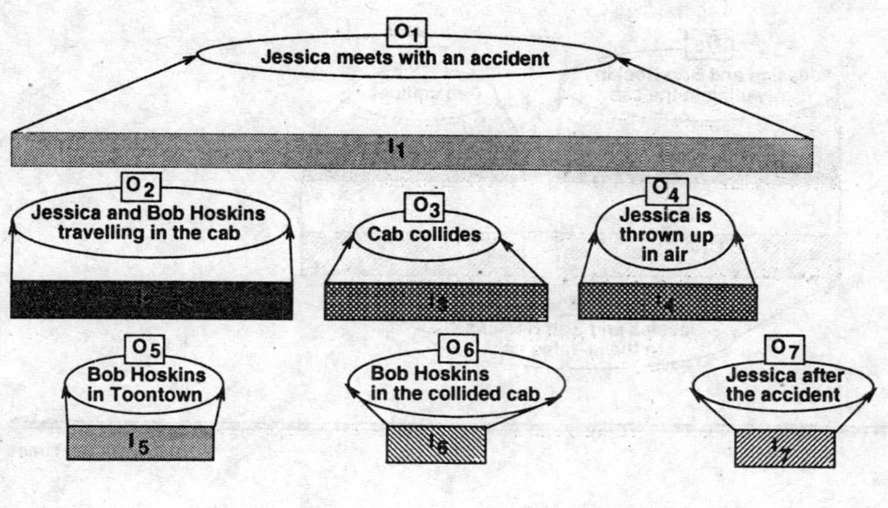

Figure 5.3 Interval Based Inheritance

- *Location* : Toontown.

Interval Based Inheritance : OVID follows an inheritance mechanism by which *some* of the attribute/value pairs of a video object are inherited by another object if the former object's intervals contains the latter object's intervals. This concept is called *interval inclusion relationship.* As an example, in Figure 5.3, the object o_3 is *included* by the object o_1. Hence, o_3 inherits some of the attributes/values of the object o_1. o_3 inherits the value of the location attribute, *Toontown*, from o_1. Some of the attribute/values cannot be inherited since their inheritance may not be meaningful. For example, the attribute *location* can be considered as an inheritable attribute whereas *event* cannot be.

Composition Operations for OVID Objects : The following operations are defined in order to compose new OVID objects from existing ones.

- Projection
- Merge

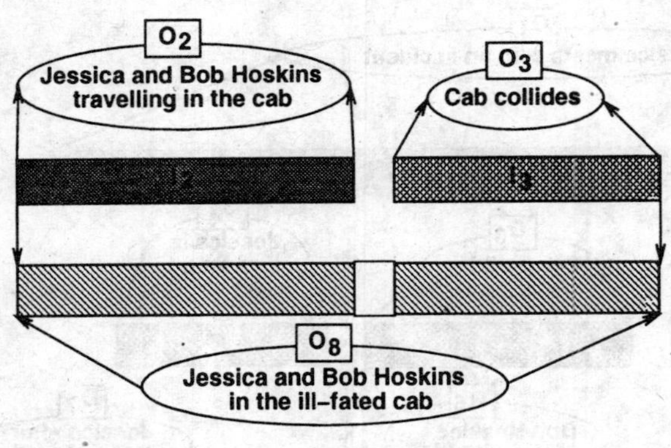

Figure 5.4 **Merging Video Objects**

- Overlap

These operations use the mechanism of interval based inheritance. *Projection* operation helps in composing a new object for a certain portion of the scene corresponding to an already existing object. In the projection operation, the new object inherits some of the attributes/values pair of the existing object. In Figure 5.3, objects o_2, o_3, o_4, o_5 and o_6 can be considered to be projected from the object o_1. All these projected objects inherit the value of location attribute, *Toontown*, from o_1.

Merge operation creates a new object from existing objects o_i and o_j such that some attributes/values common to both o_i and o_j are inherited by the new object. In Figure 5.3, we can consider merging of the objects o_2 and o_3. The result of the merge operation is shown in Figure 5.4. The attributes/values *common* to both o_2 and o_3 are inherited by o_8.

In a similar manner, *overlap* operation extracts scenes that are described by two existing objects and compose them into a new object. The new object then inherits the attributes/values pairs of both the input objects. Figure 5.5 describes the overlap operation on the objects o_4 and o_7 described in Figure 5.3. The object o_9 inherits the attributes of both o_4 and o_7.

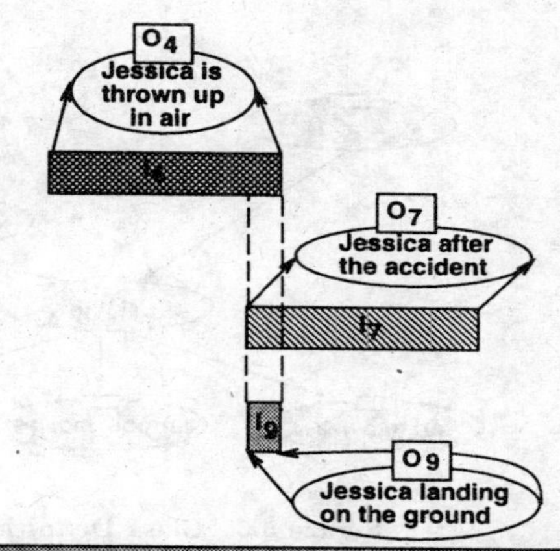

Figure 5.5 Overlapping Video Objects

Jasmine Approach

An object-oriented model, termed *Jasmine,* which includes an object model and a knowledge-base programming language has been described in [94]. The object model and the associated programming language have the following salient features :

- Set-oriented object access. This feature provides an associative access to objects. Class definition in Jasmine is the set union of the instances of a class as well as its subclasses. For instance, consider the movie class and its subclasses, shown in Figure 5.6 (this class DAG can be considered part of the one shown in Figure 5.2). The movie class represents its definition, the set of its instances as well as the instances of its subclasses, *cartoon_movies*, *art_movies* and *comedy_movies.*

- System-defined or built-in objects provide support for developing a wide variety of multimedia applications. System-defined objects have associated system-defined methods. In addition, users can develop their custom defined methods as well.

- Qualified attributes of an object, with a set of descriptions called *facets.* These facets help in describing restrictions or conditions that might be associated with an attribute.

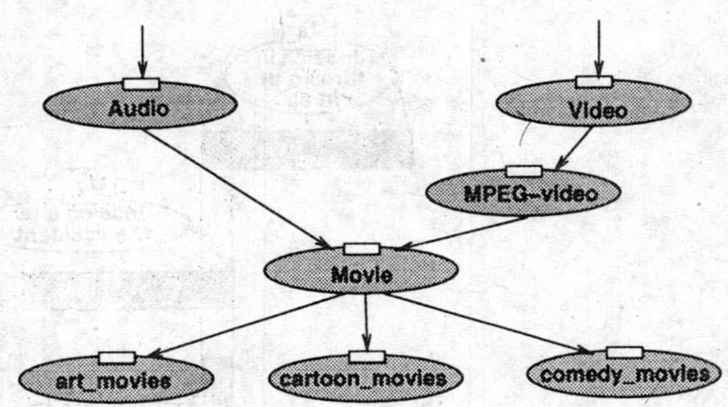

Figure 5.6 Class Definition in Jasmine

```
MOVIE
enumerated    AUDIO audio
              VIDEO MPEG_video
              STRING movie_name mandatory
              STRING direction mandatory
              FLOAT movie_play_time
              constraint { (value > 0.0 && value < 180) }
procedural    PLAY_MOVIE play_audio_video()
              {
              PLAY_MOVIE mp ;

              mp = <PLAY_MOVIE>.instantiate();
              mp.audio = self.audio.play;
              mp.video = self.video.play;
              :
              return mp;
              }
```

Table 5.1 Class Definition in Jasmine : An Example

The example of *MOVIE* class description in Jasmine is shown in Table 5.1. Here, the keyword *enumerated* describes user-supplied enumerated attributes. Each attribute has a name and belongs to a particular class. In the above example, the attribute names *movie_play_time* belongs to the class *FLOAT*. The value assigned for an attribute is *qualified* by the facets such as *mandatory* and *constraint*. The facet *mandatory* describes that the attribute *has* to be initialized to *a* value. The facet *constraint* denotes that the value of an attribute has to satisfy the specified constraints. In the above example, the value of the attribute *movie_play_time* has to be greater than 0 and less than 180 minutes. The Jasmine describes other facets such as *default* (default value to be referenced), *multiple* (multiple values for an attribute) and *common* (a value that is common to all instances of a class). Methods or procedural attributes associated with a class are specified by the keyword *procedural* (e.g., *PLAY_MOVIE*).

Knowledge-base Programming Language : Jasmine integrates a general purpose programming language (C) with features for object manipulation, calling it *Jasmine/C*. Set-oriented access feature allows objects to be operated upon in an associative manner. The objects to be operated on can be specified in the following manner :
class
class.attribute-1.attribute-2...attribute.n
The object expression *class* operates over a set of object instances belonging to a particular class as well as its subclasses. In a similar manner, the expression, *class.attribute-i* operates over a set of object instances as well as their attributes. Querying the database is expressed in the form : < *targetpart* > where < *conditionpart* > . In the Jasmine system, the *target* part consists of an object expression, or a list of object expressions. As an example, a query for playing the movie *Who Framed Roger Rabbit* will appear in the Jasmine approach as : *MOVIE.play_video_audio() where MOVIE.movie_name == "Who Framed Roger Rabbit"*.

5.1.5 Summary

An object encompasses the code that operates on its data structure. The external access interface provided to other objects is in the form of messages exchanged. Encapsulation helps in *hiding* the implementation details of the object. It also helps in system evolution since modification of an object's implementation does not necessitate changes in the code of other objects as long as the external interface remains unchanged.

Object-oriented modeling seems to be natural for most multimedia applications. Different media composing a multimedia application and the operations to be carried out on them, can be modeled using the object-oriented approach. For modeling multimedia applications, certain additional features can be provided in the object-oriented approach. These features include set-oriented object access, class hierarchy independent database schema and media specific features

Object-oriented Models	Desirable Features
	(i) Set-oriented object access (ii) Database schema independent of class hierarchy (iii) Specific features for different media objects (e.g.) interval based inheritance for video objects

Table 5.2 Desirable Features For Object-Oriented Modeling

such as interval based inheritance for video objects. Table 5.2 summarizes the desirable features for object-oriented multimedia database modeling. As case studies, we discussed OVID (Object Video Information Database) and Jasmine approaches.

5.2 TEMPORAL MODELS

The objects composing a multimedia database have an associated temporal characteristics. These characteristics specify the following parameters.

- Time instant of an object presentation.
- Duration of presentation.
- Synchronization of an object presentation with those of others.

The above parameters can be specified either in a hard or a flexible manner. In the case of hard temporal specification, the parameters such as time instants and durations of presentation of objects are fixed. In the case of flexible specification, these parameters are allowed to vary as long as they preserve certain specified relationships. As an example, consider the following temporal specifications :

- (a) Show the video of the movie *Toy Story* AT 11 am FOR 10 minutes.
- (b) Show the video of the movie *Toy Story* SOMETIME BETWEEN 10.58 am and 11.03 am, till the audio is played out.

First one, (a), is a hard temporal specification with the time instant and duration of presentation fixed at 11 am and for 10 minutes, respectively. Whereas the specification (b) is a flexible one in that it allows the presentation start time to vary within a range of 5 minutes and the duration of video presentation till the corresponding audio is played out.

The temporal specification, apart from describing the parameters for an individual object presentation, also needs to describe the synchronization among the composing objects. This synchronization description brings out the temporal dependencies among the individual object presentations. For example, in the above temporal specification (b), video has to be presented till the audio object is presented. Hence, a temporal specification needs to describe individual object presentation characteristics (time instant and duration of presentation) as well as the relationships among the composing objects. Also, users viewing multimedia data presentation can interact by operations such as fast forwarding, rewinding and freezing. The temporal models also need to describe how they handle such user interactions.

5.2.1 Modeling Temporal Relations

Given any two multimedia object presentations, the temporal requirements of one object can be related to that of another in thirteen possible ways, as shown in Figure 5.7. These thirteen relationships describe how the time instants and presentation durations of two multimedia objects are related. These relationships, however, do not quantify the temporal parameters, time instants and duration of presentations. Many models have been proposed to describe the temporal relationships among the multimedia objects. Now, we shall discuss some of these temporal models.

Hard Temporal Models

These models describe the temporal relationships in a precise manner by specifying exact values for the time instants and durations of presentations. The simplest model is the *timeline model.* In this model, media objects are placed on a timeline describing the values for the time instants and presentation durations. Figure 5.8 shows the timeline model of the VoD database example discussed in Chapter 1. For example, the values for time instant and duration of presentation of the text object W are t_1 and $t_7 - t_1$. Due to its simplicity, the timeline model has been extensively used in describing the temporal relationships in multimedia databases. However, this model describes only the

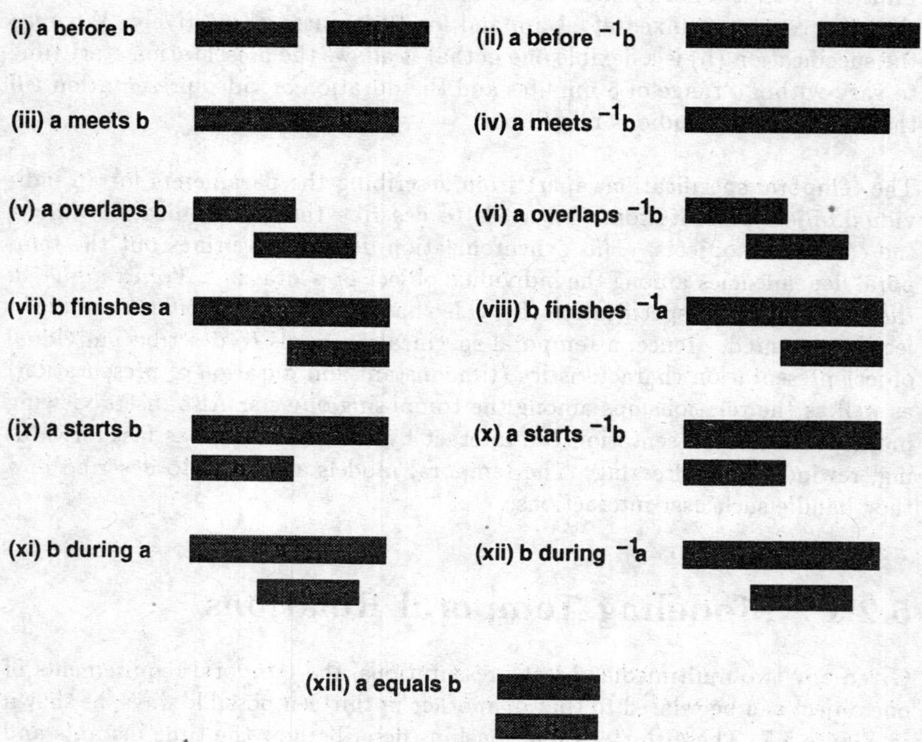

Figure 5.7 13 Possible Temporal Relations

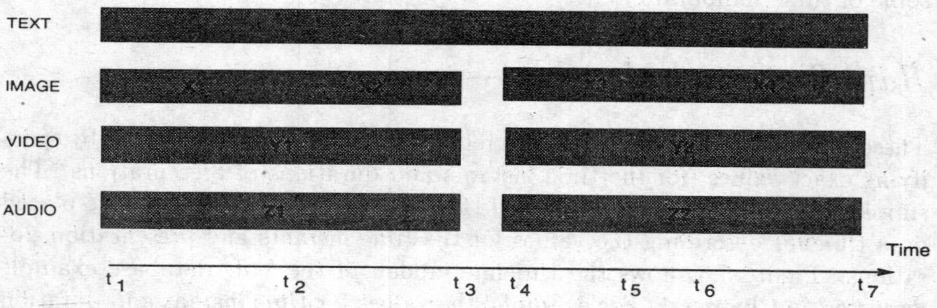

Figure 5.8 Time-line Model

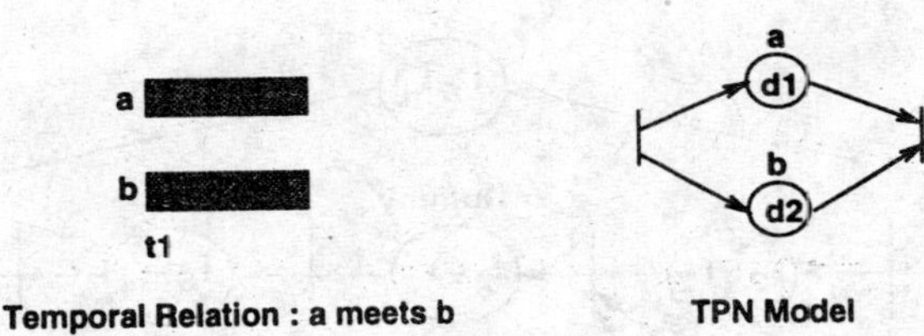

Figure 5.9 Timed Petri Nets Model

parameters for individual objects and not the presentation dependencies among the objects. For example, in Figure 5.8, video object *Y1* and audio object *Z1* have to be presented simultaneously. This dependency is not explicitly brought out in the timeline model.

Graphical Models : Graphical models have been used to describe the temporal requirements of a multimedia database. These models are based on Petri nets and Time-Flow Graphs. Petri Nets have the ability to describe real-time process requirements and interprocess timing relationships, as required for multimedia presentations. A Petri net is a bipartite graph consisting of *place* nodes and *transition* nodes. Places, represented by circles, are used to represent conditions; transitions, drawn as vertical bars, are used to represent events. For example, a place can describe the presentation of a multimedia object and a transition can represent the completion of the multimedia presentation. When representing presentation of multiple objects, the transitions can serve as a representation of synchronization characteristics of the presentation. For the purpose of modeling time-driven systems, the notion of time was introduced in Petri nets, calling them as Timed Petri Nets (TPN). In TPN models, the basic Petri net model is augmented by attaching an execution time variable to each node in the net. The time durations can be attached either to places or to transitions.

The TPN model can be used for modeling the temporal requirements of multimedia database applications. Figure 5.9 shows the TPN model for a temporal relation : object a *meeting* b. The objects have the same presentations durations, d1 = d2, and a start time, t1. The object presentations are denoted by *places* (circles) and the presentation durations are represented as values assigned to places. The transitions represent the synchronization of the start and the completion of presentation of the objects *a* and *b*. Figure 5.10 shows the TPN model for describing the synchronization characteristics of the VoD database example described in Figure 5.8.

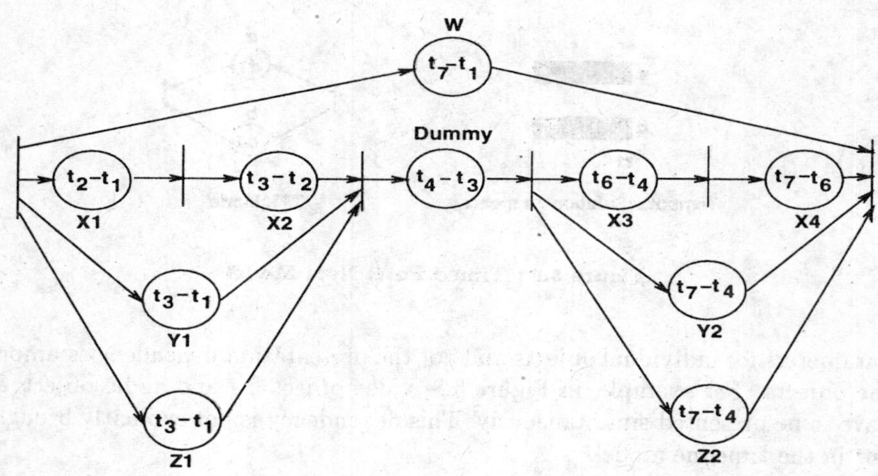

Figure 5.10 TPN Model For Figure 5.8

Flexible Temporal Models

These models represent the temporal requirements in a *soft* manner. Here, the start time, duration of presentation and the synchronization among different objects are described with a range of values (in contrast to a single value in hard temporal specification). Figure 5.11 describes a flexible temporal specification for the temporal relation : object a *before* object b. The values for the durations of presentations of the objects, $d1$ and $d2$, have ranges, $x_6 - x_5$ and $x_8 - x_7$, respectively. Similarly, the presentation start times of the objects a and b related by the range specified by the relation, $x_3 < t2 - t1 < x_4$. This type of range specification gives flexibility to the temporal parameters. Difference constraints can be used to describe this flexibility in multimedia presentation. The difference constraints specifications are similar to the value range specifications described above. However, the difference constraints specifications have a particular structure for describing the range of values. As an example, the difference constraints specification for the presentation start times $t1$ and $t2$ of objects a and b in Figure 5.11 will be represented as $t2 - t1 \geq u$ (u being a positive real number).

In a similar manner, relations between other temporal parameters can be represented as difference constraints. These difference constraints specifications have to be *solved* to *select* values for the temporal parameters. For example, a solution for the value $d0$ (in Figure 5.11) has to lie within $x1$ and $x2$. Different

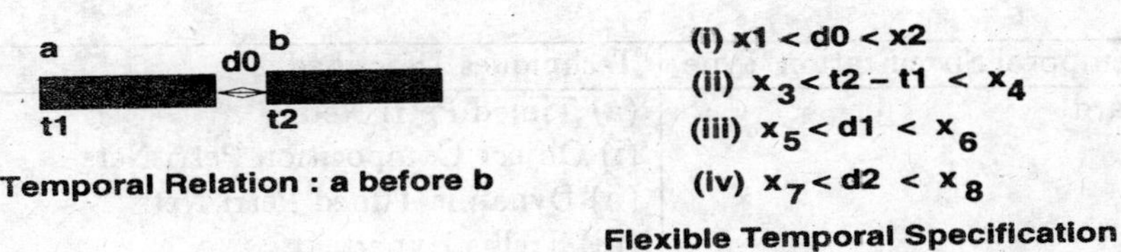

Figure 5.11 Flexible Temporal Specification

Figure 5.12 Enablers and Inhibitors in FLIPS Model

methodologies can be used for solving these constraints specification. Variations of shortest-path algorithms and linear programming approaches can be used for solving difference constraints.

A concept of *barriers* and *enablers* has been used for describing temporal requirements. This model, called *Flexible Interactive Presentation Synchronization* (FLIPS), describes the synchronization of multimedia objects using relationships between the presentation events (refer [160]). The presentation events considered by FLIPS are *Begin* and *End* of an object presentation. FLIPS employ two types of relationships, *enabling* and *inhibitive*. For example, the *End* of an object presentation can *enable* the *Begin* of another object presentation, or an object presentation can be forced to end when another object finishes. Figure 5.12 (i) shows an enabling relationship for the temporal relation, b finishes a. Here, b is forced to end when a ends. In a similar manner, the *inhibitive* relationship prevents an event from occurring until another one has occurred. Figure 5.12 (ii) describes the inhibitive relationship for the temporal relation, a before b. Here, the start of presentation of object b is inhibited till the end of a.

Temporal Specification Type	Techniques Described
Hard	(a) Timed Petri Nets (i) Object Composition Petri Nets (ii) Dynamic Timed Petri Nets (iii) Trellis Hypertext
Flexible	(a) Difference Constraints (b) Enablers/Inhibitors (FLIPS model)

Table 5.3 Techniques For Temporal Constraints Specification

5.2.2 Summary

Multimedia objects have an associated temporal specification that describes the time instants, durations, and synchronization of object presentations. The temporal specifications can be hard or soft. Hard temporal models specify exact values for the time instants and durations. We described Timed Petri nets (TPN) based models for hard temporal specification. Flexible temporal models specify a range of values for time instants and durations of presentations. We described difference constraints based approach and FLIPS model for this purpose. Table 5.3 summarizes the techniques used for temporal models.

5.3 SPATIAL MODELS

Most multimedia objects have to be delivered through windows on a monitor. Multimedia databases might include specification of the spatial layout of the various windows on the monitor. This specification might have to be done in such a way that presentation of different objects do not overlap. Figure 1.6 shows a possible spatial organization for presenting the objects in the VoD server example discussed in Chapter 1. A window on the monitor can be specified using the position (x and y coordinates) of the lower left corners and the top right corners. A window can also be specified relative to the position of another window.

The layout of the windows for presenting the objects in the example VoD database (shown in Figure 1.6) can be specified, as shown in Figure 5.13. Here, the lower left and top right corners of each window are numbered, and the corresponding x as well as y coordinates are shown. As in the case of temporal

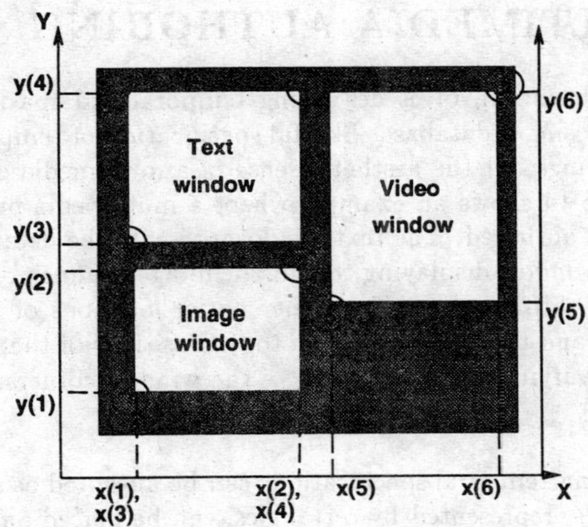

Figure 5.13 Spatial Characteristics Representation

models, the values of the x and y coordinates of the window corners can be specified in an absolute manner (hard spatial specification) or in a flexible manner. A hard spatial specification would assign values (corresponding to the pixel positions) for the x and y coordinates. For instance, the spatial characteristics of the image window can be specified as : x(1) = 10; y(1) = 15; x(2) = 100; y(2) = 105. In a flexible spatial specification, the x and y coordinates can be specified relatively. For instance, the positions of the image and video windows can be specified using difference constraints as follows.

1. $x(2) - x(1) \leq 100$ 2. $x(5) - x(1) \leq 200$
3. $x(6) - x(5) \leq 120$ 4. $y(2) - y(1) \leq 90$
5. $y(5) - y(1) \leq 100$ 6. $y(6) - y(5) \leq 200$

Here, specifications 2 and 5 describe the relative positions of the image and video windows (the difference between their x and y coordinates). Similarly, specifications 1 and 4 describe the position of the image window, specifications 3 and 6 describe the video window. Depending on the application, the position of the windows can be chosen in such a manner that the above specifications are satisfied. Though spatial specifications are simple, they help in artistic presentation of multimedia information.

5.4 MULTIMEDIA AUTHORING

Multimedia authoring involves describing temporal and spatial specifications of objects composing a database. Skillful specification of temporal and spatial specifications brings out the aesthetic sense of a multimedia database presentation. Figure 5.14 shows an example where a multimedia presentation on a missile launch is authored. The text window presents the details of the missile with the image window displaying the missile image. Launch of the missile can be animated by shifting the shifting the spatial locations of the window (by dX to the right and by dY towards the top). Positions of the windows can be described by specifying the coordinates of the window corners, as discussed in Section 5.3.

The corresponding temporal specifications can be authored as shown in Figure 5.15. The objects, represented by rectangles, can be placed on a timeline. For hard temporal specifications, the length of the rectangle specifies the presentation duration of an object. For instance, the presentation duration of text object is $t6-t1$. The values for time instants and durations of objects presentations, as derived from the timeline, can be used for generating the appropriate temporal model (Petri nets, Time Flow Graph, etc.). For flexible temporal specifications, say using difference constraints, arcs can be used to represent the relation between the object presentations. For example, in Figure 5.15, arc 1 specifies the following relation between the start of text and image presentations : $t2 - t1 \leq \delta$. Similarly, arc 2 specifies the duration of text presentation as follows : $t6 - t1 \leq \delta$. Arc 3 specifies the relation between start of image presentation and the missile launch animation as : $t3 - t2 \leq \delta$. The relations represented by the arcs can be used to generate the difference constraints specification for the multimedia database.

Graphical User Interface (GUI) based tools are required to facilitate multimedia authoring. Many commercial tools are available in the market for multimedia authoring. These tools are available in different platforms such as Microsoft Windows and Apple Macintosh. Some of the existing commercial tools are :

Multimedia Toolbook : runs on Microsoft Windows platform. Toolbook supports a *OpenScript* language for authoring. Authoring, using Toolbook, involves creation of a *book* and a book inturn consists of pages with objects placed on each page.

IconAuthor : runs on Microsoft Windows as well as on Mac operating system. An icon is a small picture that represents a function that can be

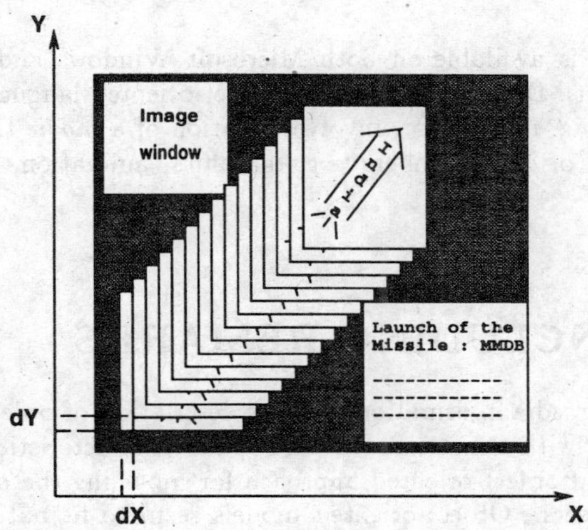

Figure 5.14 Authoring Spatial Requirements : Example

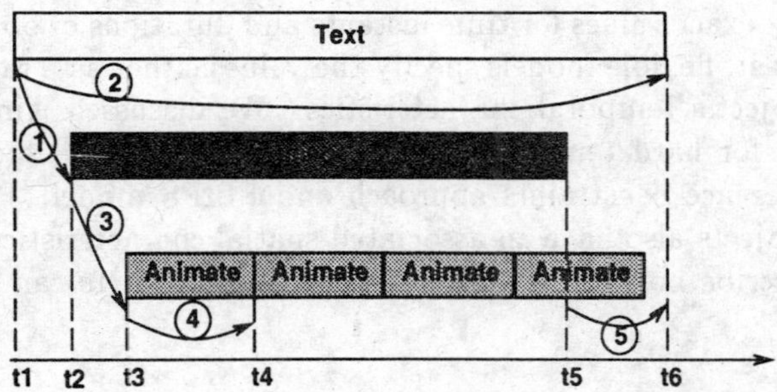

Figure 5.15 Authoring Temporal Requirements : Example

performed. Authoring involves specifying a flowchart using these icons. The flowchart describes the sequence of actions (presentation of various objects) to be performed. IconAuthor is oriented towards non-programmers.

Director : is available on both Microsoft Windows and Mac operating system platforms. Director provides an object-oriented language environment, *Lingo*. Authoring, in Director, involves creation of a *movie* that consists of a stage and a set of cast members (e.g., graphics, animation, video, text, and sound).

5.5 CONCLUDING REMARKS

Modeling multimedia information involves description of objects composing a database, along with their temporal and spatial characteristics. In this chapter, we discussed object-oriented approach for modeling the characteristics of multimedia objects. Object-oriented models seem to fit naturally for multimedia objects. Some additional features such as set-oriented object access, database schema independent of class hierarchy, and media specific features (e.g., interval-based inheritance for video objects) can be provided as part of object-oriented models for multimedia databases. OVID (Object Video Information Database) and Jasmine systems were discussed as case studies.

Temporal characteristics of multimedia objects describe the time instants, durations and synchronization of their presentation. Modeling the temporal characteristics can be done either in a hard or flexible manner. Hard temporal models specify exact values for time instants and durations of objects presentation. In contrast, flexible models specify the values either as a range or relative to another object's temporal characteristics. We discussed Timed Petri nets based models for hard temporal specification. For flexible specification, we described difference constraints approach and FLIPS model. Presentation of multimedia objects also have an associated spatial characteristics. These characteristics describe how the windows for displaying objects can be laid out on a monitor.

The temporal and spatial characteristics associated with multimedia objects have to be incorporated into their description. In order to do this, the temporal and spatial constraints have to be solved and the values have to be included in the object description. We presented authoring techniques that can be used for describing the temporal and spatial characteristics of objects. Table 5.4

Object-oriented Models	Additional Features
	(i) Set-oriented object access (ii) Database schema independent of class hierarchy (iii) Specific features for different media objects (e.g.) interval based inheritance for video objects
Temporal Specification	Techniques Described
Hard Specification	(a) Timed Petri Nets
Flexible Specification	(a) Difference Constraints (b) Enablers/Inhibitors (FLIPS model)
Spatial Specification	Difference Constraints

Table 5.4 Multimedia Information Modeling

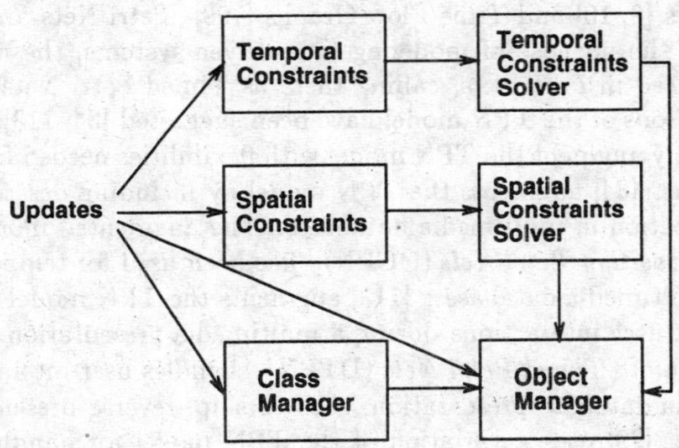

Figure 5.16 Components of Data Manager

summarizes the desirable features and the techniques used for multimedia information modeling.

Figure 5.16 shows a simple block diagram of a multimedia data manager. The class manager module maintains the hierarchy of the classes in the multimedia database. The object manager module maintains the various instantiations of the classes used. The temporal and spatial characteristics of the objects are also maintained by the object manager. The temporal characteristics are obtained from the temporal constraints solver while the spatial ones are obtained from the spatial constraints solver.

Bibliographic Notes

Modeling of multimedia information is discussed in [144, 48, 102, 94, 141]. A video database system named **OVID : O**bject **V**ideo **I**nformation **D**atabase has been introduced in [102]. An object-oriented model termed *Jasmine* which includes an object model and a knowledge-base programming language has been described in [94]. Object-oriented model of a news-on-demand server is presented in [141].

[13] presents the thirteen possible ways in which temporal requirements of two objects can be related. Graphical models have been used to describe the temporal requirements of a multimedia database [34, 113]. These models are based on Petri nets [8, 10] and Time-Flow Graphs [109]. Petri Nets are described in [8, 10]. For the purpose of modeling time-driven systems, the notion of time was introduced in Petri nets, calling them as Timed Petri Nets (TPN) [12]. Many variations of the TPN model have been suggested [34, 113]. These variations basically augment the TPN model with flexibilities needed for multimedia presentations. [34] augments the TPN model by including descriptions for resource utilization in multimedia databases. The augmented model, called the *Object Composition Petri Nets* (OCPN), has been used for temporal representation in multimedia databases. [113] augments the TPN model with facilities for handling user interactions during a multimedia presentation. This model, termed *Dynamic Timed Petri Nets* (DTPN), handles user interactions during a multimedia database presentation, such as skip, reverse presentation, freeze and resume. [29] uses a variation of the TPN model for handling hypertext applications.

[92, 163] used the concept of difference constraints to describe flexibility in multimedia presentation. [163] uses variations of shortest-path algorithms for solving the difference constraints. A concept of *barriers* and *enablers* have been used for describing temporal requirements in the FLIPS (Flexible Interactive Presentation Synchronization) model [160].

[156, 162, 163] describe the issues in multimedia authoring systems. Multimedia Toolbook is described in [169]. The features of IconAuthor are presented in [170] and those of Director can be found in [171].

6

QUERYING MULTIMEDIA DATABASES

A *query* is a language expression that describes the data to be retrieved from a database. A typical query has the following components:

- The data item(s) that is(are) desired as output
- The information base in which the search is to be made
- The conditions, termed *query predicates*, that have to be satisfied for a data item to be selected as output data

Queries on multimedia databases can be of different types based on what the predicates describe and how the predicates are specified. In this chapter, we present how different types of queries can be processed. We also examine language features for describing queries.

6.1 QUERY PROCESSING

As discussed in Section 1.4.4, multimedia queries can be of different types and they can be processed in the following manner :

1. **Query on the content of the media information :** (Example Query: Show the details of the movie where a cartoon character says : '*Somebody poisoned the water hole*').
   The content of media information is described by the metadata associated with media objects (as discussed in Chapter 3). Hence, these queries have

to be processed by accessing directly the metadata and then the media objects.

2. **Query by example (QBE)** : (Example Query : Show me the movie which contains *this* song.)

   QBEs have to be processed by finding a similar object that *matches* the one in the example. The query processor has to identify exactly the characteristics of the example object the user wants to match. We can consider the following query : Get me the images *similar* to *this* one. The similarity matching required by the user can be on texture, color, spatial characteristics (position of objects within the example image) or the shape of the objects that are present in the image. Also, the matching can be exact or partial. For partial matching, the query processor has to identify the degree of mismatch that can be tolerated.

   Then, the query processor has to apply the cluster generation function for the example media object. As discussed in Sections 4.1.4 and 4.3.2, these cluster generating functions map the example object into an m-dimensional feature space. The query processor has to identify the objects that are mapped within a distance d in the m-dimensional feature space (as shown in Figure 4.11 for image objects). Objects present within this distance d are retrieved with a certain measure of confidence and are presented as an ordered list. Here, the distance d is proportional to the degree of mismatch that can be tolerated.

3. **Time indexed queries** : (Example Query : Show me the movie *30 minutes after* its start).

   These queries are made on the temporal characteristics of the media objects. The temporal characteristics can be stored using segment index trees, as discussed in Section 4.4.1. The query processor has to process the time indexed queries by accessing the index information stored using segment trees or other similar methods.

4. **Spatial queries :** (Example Query : Show me the image where President Yelstin is seen to *the left of* President Clinton).

   These are made on the spatial characteristics associated with media objects. These spatial characteristics can be generated as metadata information. The query processor can access this metadata information (stored using techniques discussed in Section 4.3.1) to generate the response.

5. **Application specific queries** : (Example Query : Show me the video where the river *changes its course*).

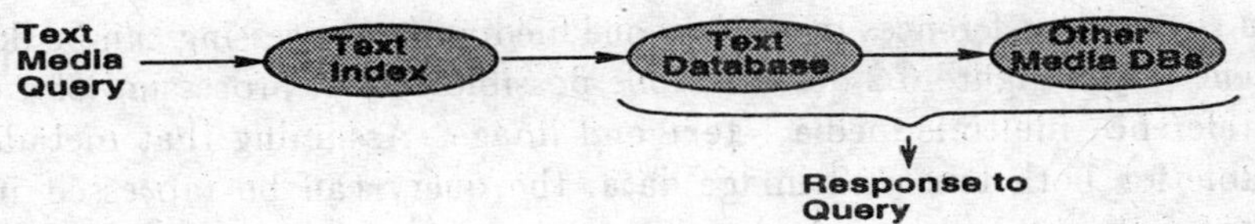

Figure 6.1 Processing Single Media Query

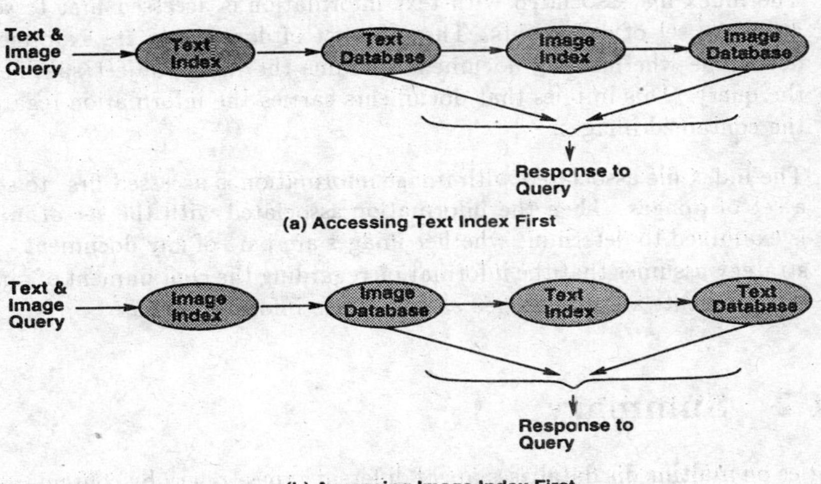

Figure 6.2 Processing Multiple Media Query

Application specific descriptions can be stored as metadata information. The query processor can access this information for generating response.

6.1.1 Options For Query Processing

Queries in multimedia databases may involve references to multiple media objects. Query processor may have different options to select the media database that is to be accessed first. As a simple case, Figure 6.1 describes the processing of a query that references a single media, *text*. Assuming the existence of metadata for the text information, the index file is accessed first. Based on the text document selected by accessing the metadata, the information is presented to the user.

When the query references more than one media, the processing can be done in different ways. Figure 6.2 describes one possible way of processing of a query that reference multiple media : text and image. Assuming that metadata is available for both text and image data, the query can be processed in two different ways :

- The index file associated with text information is accessed first to select an initial set of documents. Then this set of documents are examined to determine whether any document contains the image object specified in the query. This implies that documents carries the information regarding the contained images.
- The index file associated with image information is accessed first to select a set of images. Then the information associated with the set of images is examined to determine whether images are part of any document. This strategy assumes that the information regarding the containment of images in documents are maintained as a separate information base.

6.1.2 Summary

Queries on multimedia databases are of different types: query by content, query by example, time indexed, spatial, and application specific. Processing these different types of queries are carried out by

- Accessing metadata associated with the objects
- Applying cluster generation functions on example objects

Table 6.1 summarizes the methodologies used for processing different types of queries.

6.2 QUERY LANGUAGES

The conditions specified as part of a user query that are to be evaluated for selecting an object are termed as *query predicates*. These predicates can be combined with boolean operators such as AND, OR, and NOT. Query languages are used to describe query predicates. For multimedia database applications, query languages require features for describing the following predicates.

Query Type	Processing Methodology
Query By Content	Access metadata associated with media objects
Query By Example	Use cluster generation methodologies for the example object
Time Indexed	Access index on temporal information
Spatial	Access metadata associated with spatial data
Application Specific	Access application specific metadata

Table 6.1 Desirable Features For Multimedia Query Processing

- Temporal predicates
- Spatial predicates
- Predicates for describing queries by example
- Application specific predicates

Apart from features required for describing different predicates, query languages also require features for describing various media objects. Different query languages are used for multimedia database applications. Structured Query Language (SQL) has been defined in the seventies by IBM, for traditional databases. International Standards Organization (ISO) has been trying to standardize on different versions of SQL : SQL89, SQL2, SQL3 and SQL/MM. SQL and its derivatives do offer features for describing the multimedia database queries. However, multimedia database applications have a wide range of requirements. Hence, various research groups have proposed other query languages. Each query language offers features to facilitate description of queries for a particular category of applications. In this section, we shall discuss salient features of the following query languages that have been suggested for multimedia database applications.

- Structured Query Language for Multimedia (SQL/MM)
- PICQUERY+
- Video SQL

6.2.1 SQL/MM

SQL/MM offers new data types such as Binary Large Objects (BLOBs), new type constructors, and object-oriented features. The new built-in data types are provided as *Abstract Data Types.* The addition of object-oriented features is to make the language more suitable for multimedia database applications. SQL/MM, as per the current status of its definition, consists of three parts : framework, full-text, and spatial part. Other parts for handling audio, video, and images are currently being worked on. We shall first discuss the *Abstract Data Type*, defined as part of SQL/MM.

Abstract Data Types in SQL/MM

The concept of abstract data type in the definition of SQL/MM allows definition of data types according to the needs of the application. This concept of ADT is similar to the definition of objects in object-oriented systems. The ADT definition has two parts : structural and behavioral. The structural part defines the data structures that are part of the ADT and the behavioral part describes the operations that are to be carried out on the data. Every ADT has a built-in *constructor* function defined as part of its behavioral part. The constructor function initializes the various data structures defined in the structural part. Every ADT also has a built-in destructor function that is invoked to *clean up* when the ADT is destroyed. An ADT can be defined as shown in the following example :

```
CREATE VALUE TYPE Stack {
PUBLIC x REAL(50), top INTEGER, bottom INTEGER,


PUBLIC FUNCTION – 'constructor'
m-stack () RETURNS Stack
BEGIN
DECLARE temp Stack;
SET temp = Stack() ; – set with NULLs
SET temp..top = 0;
```

```
SET temp..bottom = 0;
END ;

PUBLIC FUNCTION – Push Operation
push(x, value) .....

PUBLIC FUNCTION – Pop Operation
pop(x, value) .....
}
```

The above ADT definition describes a STACK. The structural part of the ADT consists of the variables *x, top* and *bottom. m-stack* is the user-defined constructor function that helps in initializing the defined data structures. *m-stack* calls the built-in constructor function *Stack* that initializes the local variable *temp.* Then the *top* and *bottom* pointers are initialized to 0. The behavioral part of the ADT consists of the functions *push* and *pop.* The keyword PUBLIC describes the access level (or the encapsulation level) of a variable or a function. PUBLIC description implies that the variable and the function can be accessed and called from outside the ADT. The definitions for access levels follow the normal object-oriented concepts.

Subtyping : For describing derived objects, the UNDER clause is used as follows : CREATE OBJECT TYPE obj1 UNDER obj. This declaration states that the object *obj1* is a subtype of *obj* and, the other way around, *obj* is a supertype of *obj1.* A subtype inherits all the data structures and the associated functions defined as part of its supertype. In addition, the declaration can specify data structures and functions that are to be used only within the subtype. Subtype declaration can lead to a hierarchy of objects, in a similar manner to the concept of inheritance discussed in Section 5.1.

Subtyping in SQL/MM also supports the following properties that are normally used in object-oriented languages.

- **Substitutability :** refers to using an instance of a subtype instead of its supertype.
- **Functions overloading :** implies that a function declared in the supertype can be redefined in its subtype.
- **Dynamic binding :** An object hierarchy can result in declaration of more than one function with the same name. In this case, the selection of the appropriate function to be used for execution will be determined at the

run-time depending on the best match for the arguments. This process is referred to as dynamic binding.

SQL/MM Features

SQL/MM incorporates some multimedia packages, such as *the Framework, the FullText,* and *spatial data.*

Framework : SQL/MM offers the possibility of adding custom-made functions to built-in data types. SQL/MM uses this feature to create ADTs and functions that have general purpose applicability in a number of application areas, termed *the Framework.* As an example, the Framework includes a library of numerical functions for complex numbers. The complex number ADT includes functions such as *equals, adds, negate* and *RealPart.*

FullText : SQL/MM offers an ADT termed FullText that has a built-in function, called *Contains.* The function *Contains* can be used to search documents. The FullText ADT has the following syntax :

```
CREATE OBJECT TYPE FullText
{
:
FUNCTION Contains (text FullText, search_expr CHARACTER VARYING
(max_pattern_length))
RETURNS Boolean
BEGIN ..... END
}
```

The function *Contains* searches a specific document with the string specified in *search_expr. Contains* can employ different types of searching methods such as wild cards, proximity indicators (e.g., the words 'multimedia' must be followed by the word 'application'). Logical operators such as OR, AND, and NOT can be used to compose more complex search expressions. The search operation uses the metadata defined for the text document (as discussed in Chapter 3). In addition, it can also use weighted retrieval techniques to improve the search efficiency.

Spatial Data : Several ADTs are defined in order to support spatial data structures. These ADTs help in handling image objects, especially in geographical applications.

Movie Information Database : An Example

The class hierarchy of the VoD database example discussed in Chapter 1 is shown in Figure 5.2. Here, four types of objects : *Text, Audio, Image* and *Video*, are defined. Functions for manipulating the information contained in the objects are defined as parts of the objects. The *Movie* class has functions defined for displaying the various media objects : *present_text, present_audio, present_image*, and *present_video*. The search function *Contains*, defined in SQL/MM, can be used to locate information in media objects. The VoD example discussed in Chapter 1, can be described using SQL/MM in the following manner.

```
CREATE OBJECT TYPE Text {     FUNCTION    present_text .... , }

CREATE OBJECT TYPE Audio {    FUNCTION    present_audio .... , }

CREATE OBJECT TYPE Image {    FUNCTION    present_image .... , }

CREATE OBJECT TYPE Video {    FUNCTION    present_video .... }

CREATE OBJECT TYPE Movie
{
title     CHAR(25),
info      Text,
sound     Audio,
stills    Image,
frames    Video,

FUNCTION     present_movie_info ....
}
```

Based on the object definitions for the movie information database, Query 1 discussed in Chapter 1 on the movie information database can be specified using SQL/MM as follows.

Query 1 : Give information on available movies with computerized animation cartoons?

SQL/MM Query :

```
SELECT m.title FROM Movie m
WHERE Contains (m.info, 'Computerized animation cartoons')
```

6.2.2 PICQUERY$^+$ Query Language

A query language PICQUERY$^+$ for pictorial and alphanumeric database management systems has been described in [103]. The main emphasis of the database is for medical applications. The important characteristics of medical database applications includes the following:

1. Evolutionary features : These features of a medical database describe how certain organs of a body evolve over a period of time.
   - Evolution : The characteristics of an object may evolve in time.
   - Fusion : An object may fuse with other objects to form a new object that has different features from its parent objects.
   - Fission : An object may split into two or more independent objects.
2. Temporal features : These features describe the following characteristics of the database objects.
   - Temporal relationships between two objects (e.g., an event *following* another event).
   - Time period of the existence of an object or the time point of the occurrence of an event.

PICQUERY$^+$ offers the following query operators :

- Evolutionary predicates specify the constraints associated with the different development phases of an object. The evolutionary operators, defined as part of PICQUERY$^+$, include : *EVOLVES_INTO*, *FUSES_INTO*, and *SPLITS_INTO*.
- For temporal predicates, the PICQUERY$^+$ specifies the following operators : *AFTER, BEFORE, DURING, BETWEEN, IN, OVERLAPS, MEETS, EQUIVALENT, ADJACENT, FOLLOWS*, and *PRECEDES*.
- For describing queries that deal with spatial nature of the data, the following operators are included : *INTERSECTS, CONTAINS, IS COLLINEAR WITH, INFILTRATES, LEFT OF, RIGHT OF, ABOVE, BELOW, IN FRONT OF*, and *BEHIND*.
- For describing *fuzzy* queries, operator *SIMILAR TO* is defined.

6.2.3 Video SQL

A query language, *Video SQL,* has been used in the *Object-oriented Video Information Database* (OVID) (discussed in Section 5.1.4) [102]. Video SQL is oriented towards facilitating retrieval of video objects in the OVID system. The language definition of Video SQL has the following clauses :

- **SELECT** clause, as defined in Video SQL, is different from ordinary SQL definition. It specifies the type of the OVID object that is to be retrieved: *continuous, incontinuous*, and *any*. *Continuous* denotes video objects comprising of a single sequence of frames. *Incontinuous* describes video objects consisting of more than one sequence of frames. For example, an object can consist of the frames : (1,10) and (15,30). The intermediate frames (11,14) are not considered as part of this example OVID object. *Any* describes both the categories.
- **FROM** clause specifies the name of the video database.
- **WHERE** clause describes a condition, consisting of attribute/value pairs and comparison operators. Video frame number can also be specified as part of a condition. A condition can be specified as follows :
  - *[attribute] is [value | video object]*. Here, the condition describes video objects that have the specified attribute values or video-object.
  - *[attribute] contains [value | video object]*. This condition describes the video objects that contain the specified value in a set of attributes.
  - *definedOver [video sequence | video frame]*. This condition denotes the video objects that are defined over the specified video sequences or frame.

6.2.4 Summary

Query languages for multimedia database applications require features for describing the characteristics of media objects as well as different types of query predicates. Since multimedia databases are highly application specific, application specific query languages are also used. We described the features offered by query languages such as SQL/MM, PICQUERY$^+$, and Video SQL. Table 6.2 summarizes the features of the query languages discussed so far.

Query Language Features	Example
(i) Ability to represent new data types	Abstract Data Types in SQL/MM
(ii) Temporal predicates	As in PICQUERY$^+$
(iii) Spatial predicates	— do —
(iv) Application specific predicates	— do —
(v) Media specific predicates	As in Video SQL

Table 6.2 Desirable Features For Multimedia Query Language

6.3 CONCLUDING REMARKS

Querying multimedia databases can be in different ways, such as querying by content, querying by example, time indexed queries, spatial queries, and application specific queries. The queries can be processed by accessing the metadata or by applying cluster generation functions on the example objects (for processing query by example).

Languages used for describing multimedia database queries require features for specifying different types of predicates such as, temporal, spatial, application specific, and query by example. In this chapter, we described the features of query languages such as SQL/MM, PICQUERY$^+$, and VideoSQL. Table 6.3 summarizes the methodologies used for processing different types of queries and the features of the query languages discussed.

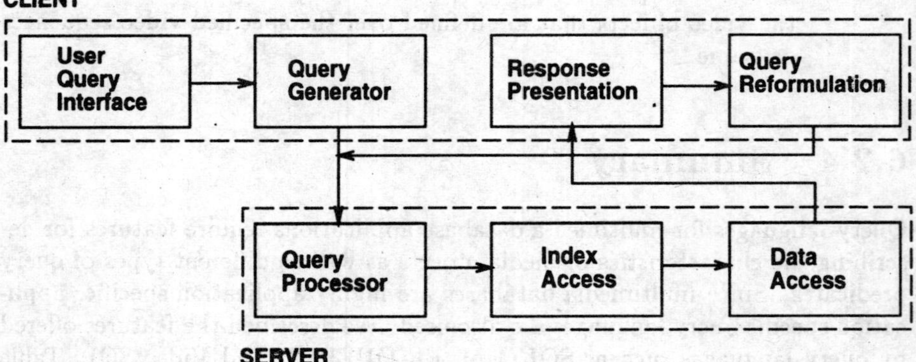

Figure 6.3 Components of Query Manager

Figure 6.3 shows a simple block diagram of query manager. The user query interface module helps a user to describe a query. The query generator module generates an appropriate query which is handled by the query processor module. The query processor accesses the required metadata as well as the objects and generates the response. The response presentation module presents the response to the user. If the response is not satisfactory, the query reformulation module helps in reformulating the user's query. In a distributed environment, a client formulates the query and handles the response from the server using the following modules : user query interface, query generator, response handler, and query reformulator. The server receives and processes a client's query using the modules : query processor, index access and data access.

Bibliographic Notes

[28, 146, 148] discusses the various issues in multimedia query processing. [43] describes the query processing in MULTOS office filing system. It also provides a classification of query predicates. The MULTOS query language has been introduced in [42]. Retrieval of multimedia documents are discussed in [22, 28, 44].

SQL, SQL/MM and their applications for multimedia database querying are introduced in [142, 155]. A query language PICQUERY$^+$ for pictorial and alphanumeric database management systems has been described in [103]. A query language, *Video SQL,* has been used in the *Object-oriented Video Information Database* (OVID) [102].

Query Type	Processing Methodology
Query By Content	Access metadata associated with media objects
Query By Example	Use cluster generation methodologies for the example object
Time Indexed	Access index on temporal information
Spatial	Access metadata associated with spatial information
Application Specific	Access application specific metadata
Query Language Features	**Example**
(i) Ability to represent new data types	Abstract Data Types as in SQL/MM
(ii) Temporal predicates	As in PICQUERY$^+$
(iii) Spatial predicates	— do —
(iv) Application specific predicates	— do —
(v) Media specific predicates	As in Video SQL

Table 6.3 Desirable Features For Multimedia Querying

7

MULTIMEDIA COMMUNICATION

Objects composing multimedia databases can be distributed over computer networks. Accessing distributed multimedia objects necessitates support from the network service providers. Large sizes of media objects influence the bandwidth (or the throughput) required for communicating these objects. Real-time nature of the media objects necessitates guaranteed delivery of objects at specified time instants.

As discussed in Section 5.2, the objects composing a response to a query have to be presented according to the temporal characteristics specified in the database. Hence, a client accessing a multimedia database server has to retrieve the objects in such a way that they can be presented according to the specified temporal schedule. In other words, the client has to determine a retrieval schedule that specifies when it should make a request for an object from the multimedia database server. This retrieval schedule depends on the throughput offered by the computer network.

In this chapter, we identify the possible ways in which a retrieval schedule can be generated, and then we examine the communication requirements of multimedia database applications.

7.1 RETRIEVAL SCHEDULE GENERATION

In a distributed multimedia database application, objects composing a database can be dispersed over computer networks. A client composes a query and com-

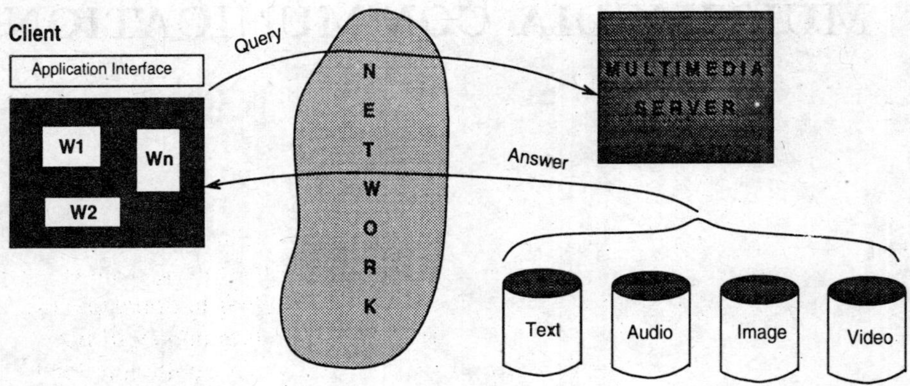

Figure 7.1 Server-Client Interaction : Single Connection

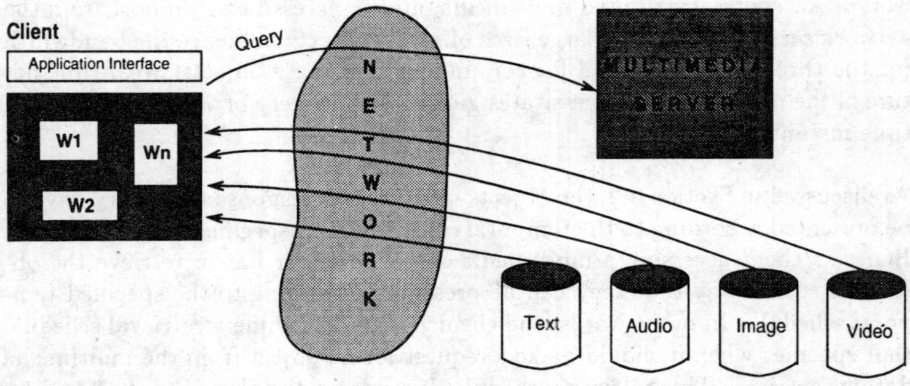

Figure 7.2 Server-Client Interaction : Multiple Connections

municates it to the server. The server processes the query, formulates the response and communicates it back to the client. This interaction between server and client is carried over communication channel(s) (also called *network connections*) established between them. Client-server interaction can be carried over a single communication channel, as shown in Figure 7.1. Here, all the media objects composing the response have to be communicated over the same channel. Alternately, multiple channels can be used for communicating individual media objects, as shown in Figure 7.2. In the case where objects are distributed on different servers, a communication channel might be required between the client and each of the servers, as shown in Figure 7.3.

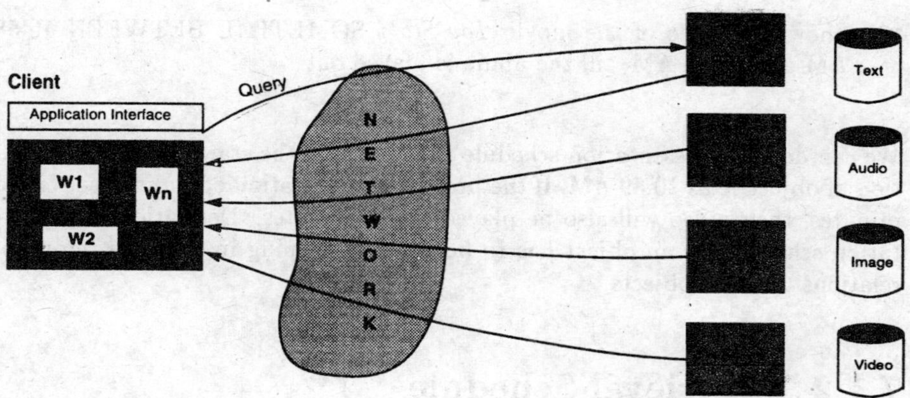

Figure 7.3 Servers-Client Interaction

The objects composing the response to the query, have to be retrieved from their server(s) and presented to the user. With the storage place acting as a *server* and the retrieving system as a *client*, the retrieval process is initiated by the client (as opposed to the server just *delivering* the objects following some schedule of its own). Hence, this retrieval process is composed of the following phases :

- Identify a *presentation schedule* that satisfies the temporal specifications associated with the multimedia database
- Identify a *retrieval schedule* that specifies the time instants at which the client should make a request to the server(s) for delivering the objects that compose the response

7.1.1 Presentation Schedule

Specification of temporal relationships among objects composing a multimedia database was discussed in Section 5.2. These relationships have to be *translated* into a presentation schedule that specifies the time instants and durations of object presentations. For hard temporal models, this presentation schedule is the same as that of the specification. However, for flexible temporal models, the specification has to be *solved* to determine a presentation schedule. As an example, consider the following temporal specification discussed in Section 5.10.

- Show the video of the movie *Toy Story* SOMETIME BETWEEN 10.58 AM and 11.03 AM, till the audio is played out

We can derive a presentation schedule that specifies the start time of presentation of object A as 10.59 AM. If the presentation duration of audio object is 15 minutes, then video will also be played for 15 minutes. Derivation of presentation schedule for an object has to be done by keeping in mind its temporal relations to other objects.

7.1.2 Retrieval Schedule

The retrieval schedule specifies when a client should retrieve an object from its server so that the object can be delivered according to its presentation schedule. As an example, in the above temporal specification for the movie *Toy Story*, if we know that the delay for retrieving the video object from its server is 3 minutes, then the retrieval schedule can be fixed at 10.56 AM (so that the movie presentation can be started according to its presentation schedule at 10.59 AM). The derivation of retrieval schedule is constrained by the following factors :

- Throughput (or the bandwidth) of the communication channel between the server and the client (i.e., the amount of data that will be sent through the network per unit time. It is specified in terms of bits per second.)
- Buffer availability in the client for retrieved objects
- Size(s) of the object(s), in terms of bytes, that is(are) to be retrieved from the server
- Time duration available for retrieval

In the above set of constraints, the throughput of the communication channel and the buffer resources are system dependent. The available throughput can vary depending on the type of network and the load on it. The buffer resources are dependent on their availability in the client system. The last two constraints : sizes of the objects and the time available for retrieval, are application dependent. The sizes of the objects depend on the type of media as well as the desired *quality* of presentation. For example, an image object may be retrieved as a thumbnail sketch or as a full image. The time available for

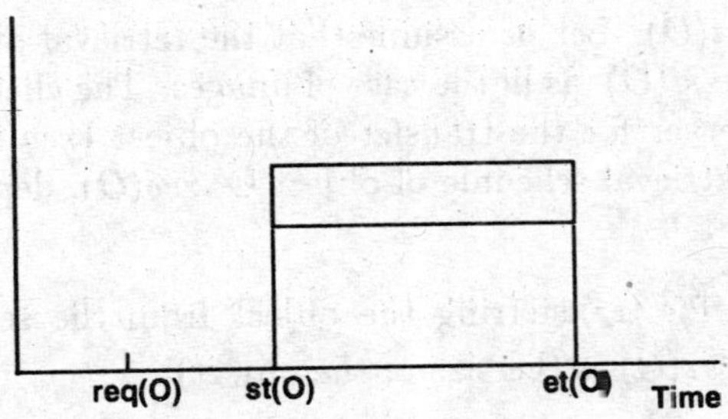

Figure 7.4 Single Object Retrieval

presentation depends on the derived presentation schedule from the temporal specification. While deriving the retrieval schedule, the following issues might have to be kept in mind.

1. Multiple objects can be retrieved over either :
   - The same network channel (as shown in Figure 7.1)
   - Different network channels (as shown in Figures 7.2 and 7.3)
2. Network provides a maximum throughput Th_{max} for each channel. Hence, this available throughput has to be shared by different objects in case their retrieval from a server has to be done simultaneously over the same channel. This throughput offered by the network service provider can vary with time, depending on the network load.
3. The client provides a maximum buffer Buf_{max} to the multimedia application for storing the retrieved objects before their presentation.
4. Depending on the media type, objects might have to be either completely or partially retrieved before the start of their presentation. For example, objects such as images have to be completely retrieved before their presentation. For objects such as video, a chunk of frames may be retrieved before the start of the entire video presentation. The rest of the frames can be retrieved as the presentation progresses.

Based on these assumptions, we now discuss how a retrieval schedule can possibly be determined.

Single Object Retrieval : As the simplest case, let us consider the retrieval of a single object as shown in Figure 7.4. The object O has to be presented by the client at time $st(O)$. Let us assume that the retrieval of the object has to be completed before $st(O)$, as in the case of images. The client makes a request at $req(O)$ to the server for the transfer of the object ($req(O)$ must be *before* $st(O)$). Here, the retrieval schedule of object O, $req(O)$, depends on :

- Time required for transferring the object from the server to the client ($\frac{sz(O)}{Th_{max}}$, where $sz(O)$ is the size of the object)
- Round trip time required for sending the request to the server and receiving the response (Δt)

Hence, for retrieval of objects such as images (whose retrieval needs to be completed before their presentation), $req(O)$ can be defined as : $req(O) = st(O) - \{\frac{sz(O)}{Th_{max}} + \Delta t\}$.

For objects such as video, $sz(O)$ can represent the chunk of frames that needs to be retrieved before the start of presentation (since whole video objects might require large buffer spaces). In the case where multiple objects are to be retrieved, the above procedure can be used if multiple communication channels are used for transferring them (i.e., one channel is used for transferring one object at a time).

Simultaneous, Multiple Object Retrieval : In many cases, multiple objects might have to be retrieved simultaneously over the same communication channel. In this case, the available throughput and the buffer resources have to be shared among the objects to be retrieved. The constraints that must be obeyed by the objects sharing the same communication path are the following:

- **Throughput :** Each communication channel has a maximum bandwidth, Th_{max}. Hence, the sum of the throughputs required by the objects that share the same path should be less than this maximum value. If objects $o_1 \dots o_n$ with throughput requirements $th(o_1) \dots th(o_n)$ are to be retrieved simultaneously over the same network channel, then $th(o_1) + \dots + th(o_n) \leq \text{Th}_{max}$.
- **Buffer :** Let the client's maximum available buffer space be Buf_{max}. If objects $o_1 \dots o_n$ are retrieved over the same network channel simultaneously and $buf(o, t)$ denotes the buffer usage of object o at time t, then $buf(o_1, t) + \dots + buf(o_n, t) \leq \text{Buf}_{max}$.

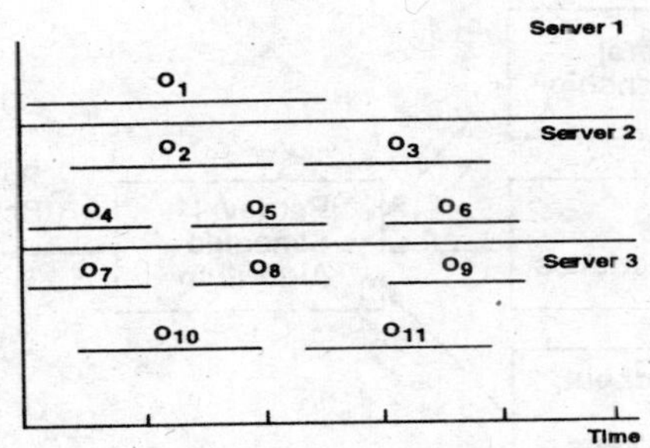

Figure 7.5 Simultaneous, Multiple Object Retrieval From Multiple Servers

Simultaneous, Multiple Object Retrieval From Multiple Servers: Figure 7.5 shows how objects composing a multimedia database presentation are retrieved from different servers. Here, separate network connections will be used for retrieving objects from the servers. Throughput constraints for multiple object retrieval have to be satisfied for each network connection separately. However, buffer constraint is the same as the entire retrieval is handled by the client.

7.1.3 Summary

In distributed multimedia database applications, a client issues a query. The response from the server(s) may be composed of multimedia objects. These objects have an associated temporal characteristics (as discussed in Section 5.2). Based on these temporal characteristics, a presentation schedule for presenting the objects has to be derived. The client has to retrieve objects composing the response from the servers so that this derived presentation schedule can be satisfied. This retrieval schedule depends on the following:

- Presentation schedule
- Sizes of the objects
- Throughput offered for the communication channel(s)
- Buffer available at the client

Figure 7.6 shows the block diagram of a simple retrieval schedule generator. The retrieval schedule algorithm takes as input the temporal relationships and

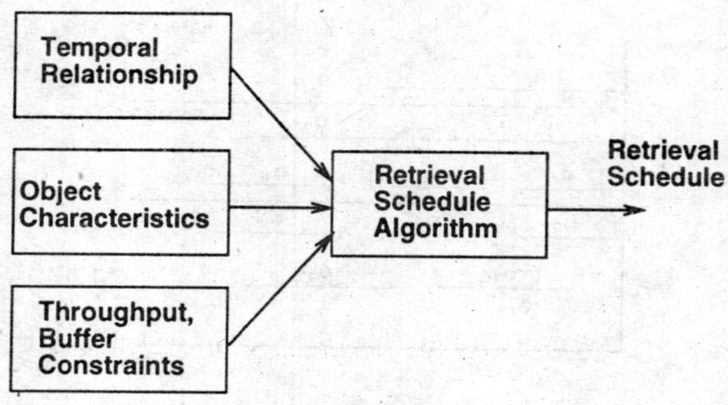

Figure 7.6 **Components of Retrieval Schedule Generator**

the object characteristics (the size of the object, whether the object has to be retrieved in full as in the case of images or in parts as in the case of video). Based on the system constraints such as throughput and buffer availability, the retrieval schedule algorithm computes a retrieval schedule.

7.2 MULTIMEDIA SERVER-CLIENT INTERACTION

We discussed the issues in generating retrieval schedules so far. The retrieval of multimedia information can be done using a single communication channel or by using multiple channels, as shown in Figures 7.1 and 7.2 respectively. Group of multiple network channels are termed as *channel groups.* Multimedia information, in response to client's query, can also be retrieved from multiple servers, as shown in Figure 7.3. Depending on the type of media information that is retrieved, the requirements of the channels might vary. These requirements are characterized by the following set of parameters, termed the Quality of Service (QoS) parameters.

1. *Traffic throughput* : This QoS parameter is related to the amount of data that will be sent through the network per unit time, specifying the traffic communication needs in terms of the bandwidth required.
2. *Transmission delay* : This parameter specifies the delay that a transmitted data unit can suffer through the network. This parameter may be expressed either in terms of an absolute or a probabilistic bound. In ad-

dition to a delay bound, a bound on the delay variation, called the ***delay jitter***, can also be specified.

3. *Transmission reliability*: This parameter is primarily related to the buffering mechanisms involved in data transmission along the network. Because of the limited size of these buffers, some packets might be lost due to traffic congestion. A probabilistic bound on such losses influence the amount of resources required for the establishment of a communication channel.

4. *Channel group relationship* : Another consideration for multimedia applications is that a group of channels might be required simultaneously for transferring different media objects. In some cases, synchronization has to be provided among channels when they are used for transferring media such as audio and video. Relationship among the channels can be specified in terms of inter-channel bounds on QoS parameters (bounds on delay jitter, in the case of audio and video channels).

Communication between a multimedia server and a client comprises of the following phases, as shown in Figure 7.7.

- Channel establishment
- Data transfer
- Channel release

Channel Establishment Phase : During this phase, the client specifies the type of QoS needed for the communication channel to the multimedia database server. The specification of the QoS parameters have to be agreed upon by the client, the server, and the network service provider. This tripartite agreement implies that sufficient resources have to be reserved by the client, the server and the network service provider in order to provide the required QoS. This process of reaching an agreement for the required QoS is termed as *QoS negotiation*. Group of channels, if required, need to be established during the connection establishment phase.

Data Transfer Phase : involves communication of multimedia information between the server and the client, as shown in Figure 7.7. The rate of data transfer should follow the throughput agreed upon during the connection establishment phase.

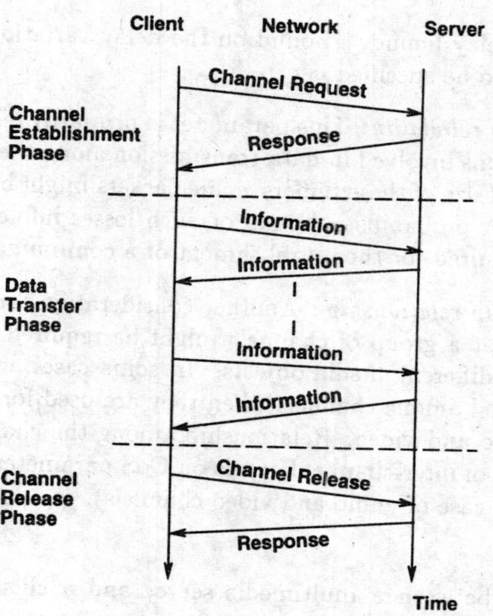

Figure 7.7 Phases in Multimedia Server-Client Communication

Channel Release Phase : involves the release of the resources held by the client, the server, and the network service provider.

The above phases are true for any server-client communication. However, for multimedia database applications, the following issues have to be addressed by the network service provider:

- QoS Negotiation
- Channel group services
- Synchronization of object transfer

7.2.1 QoS Negotiation

For demanding QoS support from the network service provider, a multimedia client should first determine the following QoS requirements.

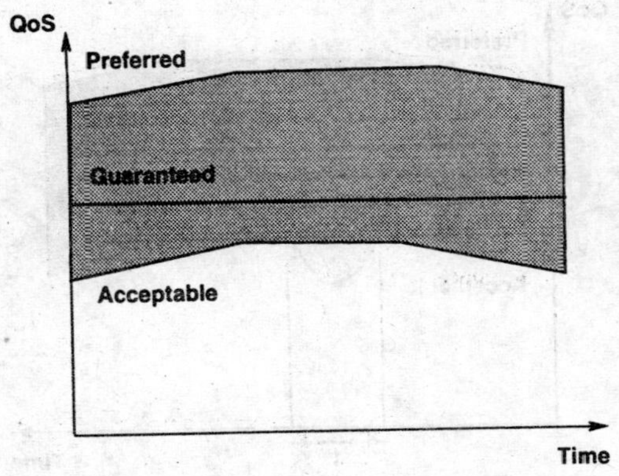

Figure 7.8 QoS For A Communication Channel

- **Preferred QoS values** : These refer to the ideal conditions for the application, say with respect to the buffering required as discussed in Section 7.1.2.
- **Acceptable QoS values** : These refer to the minimum values that are required for carrying on with the application.

Once the client determines the required QoS parameters, it has to interact with the network service provider for establishing communication channels with the required QoS parameters. The client initially makes a request for the *preferred* QoS. The network as well as the multimedia server (to which the communication channel is requested), depending on the existing load conditions, can provide the requested parameters or offer a **possible** set of QoS parameters. In the case of network offering a *possible* set of QoS parameters, the client should check the *possible* QoS with the *acceptable* values to determine whether it can tolerate the modification. If the modification is acceptable to the client, the network can establish the communication channels thereby enabling the client to carry on the communication with the server. The preferred and acceptable values denote the maximum and minimum values of the QoS spectrum, as shown in Figure 7.8. The guaranteed QoS, arrived at after negotiation with the network service provider, will be somewhere in this spectrum.

Dynamic QoS Modification : The QoS guarantees can be soft or hard. In the case of hard guarantees, the network service provider offers *deterministic* values for the QoS parameters. Here, the network cannot modify the offered QoS afterwards. In the case of soft guarantees, the network service provider of-

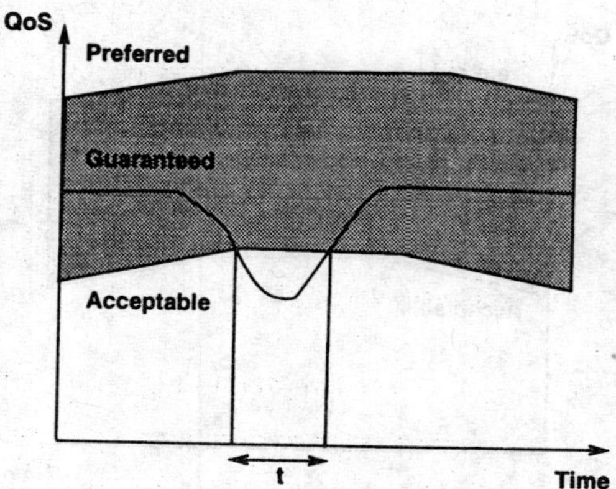

Figure 7.9 **Dynamic QoS Modification By Network Service Provider**

fers probabilistic values. If the guarantees are soft, the network service provider may modify the offered QoS parameters dynamically, depending on the load. The client should then be able to handle the dynamic modification of the QoS parameters. Figure 7.9 shows an example where modification of QoS is made dynamically by the network service provider. During the time interval t, the guaranteed QoS falls below the acceptable limit. When the modification falls within the *safe* range (between the preferred and acceptable QoS), the client can proceed smoothly. Otherwise, the application has to use other options for continuing the presentation, such as employing more buffers, slowing down the speed of presentation or dropping a media object. Some of these options can be employed only with the concurrence of the user. In the case of a dynamic modification, the client may try to *re-negotiate* its QoS requirements.

Network support for handling QoS requirements involves the following :

- Specification of the QoS required for a communication channel to a server, by a client
- Negotiation of QoS with the server

7.2.2 Channel Group Services

When media objects are retrieved over separate channels, the channels might have to be treated as a *group*. QoS negotiation and connection establishment

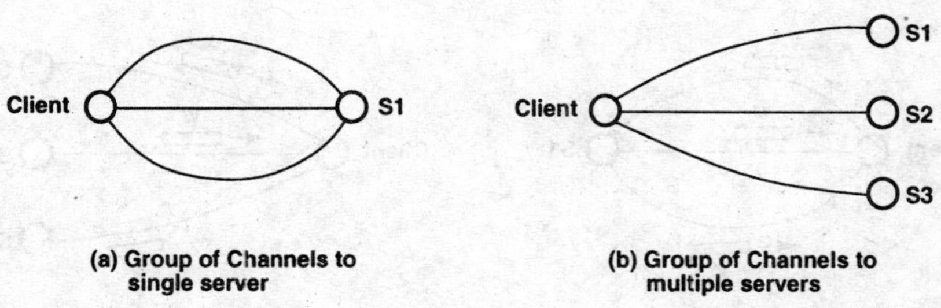

Figure 7.10 Group of Channels : Example

might have to be done for the group as a whole. This treatment of channels as a group might be necessary due to the following reasons. If one or more channels in the group cannot be established (due to any reason), then multimedia information retrieved over other channels may not make much sense. Also, objects retrieved over different channels might be related (as in the case of audio and video objects which have to be presented simultaneously).

Hence, channel group services are needed by most multimedia database applications for establishing a group of channels between server(s) and client. Group of channels can be established between a client and a server as shown in Figure 7.10(a), if all the required objects are available in the same server. Alternatively, group of channels can be established between a client and multiple servers, as shown in Figure 7.10(b). Network support for a group of channels has to be provided in terms of the following factors.

1. Creation of channel groups. This involves the following :
   - Specification of the number of channels to be established
   - Server(s) to which the channels are to be established
   - QoS required for each channel
   - Relationships, if any, among the channels
2. Joining a channel group. This feature allows a new communication channel to join an existing group.
3. Leaving a group. This feature allows a channel to be removed from an existing group.

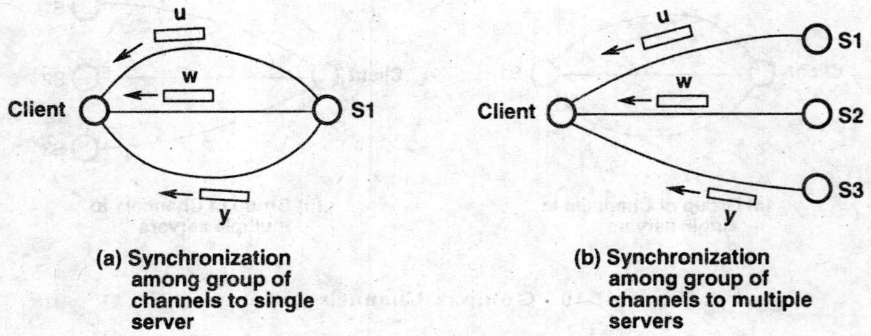

Figure 7.11 Synchronization Among Group of Channels

7.2.3 Synchronization of Object Transfer

When a group of channels are used, the transfer of media objects in different channels might have to be synchronized. For instance, video objects have to be delivered at the client along with the corresponding audio objects. Hence, the transfer of audio and video objects have to be synchronized. Figure 7.11 shows transfer of objects over channel groups. If we assume that objects u, w and y are to be delivered simultaneously at the client system, then the transfer of these objects has to be synchronized. If these objects are stored in different servers (as shown in Figure 7.11(b)), then synchronization of object transfer involves coordination of all the participating systems (client and the servers : S1, S2 and S3). Network support for synchronization of objects transfer involves :

- Specification of the objects to be transferred and the channels over which their transfer are to be synchronized

7.3 NETWORK SUPPORT FOR MULTIMEDIA COMMUNICATION

Figure 7.12 shows the components involved in providing network support to multimedia applications. The network support for multimedia applications needs to be provided in terms of the following.

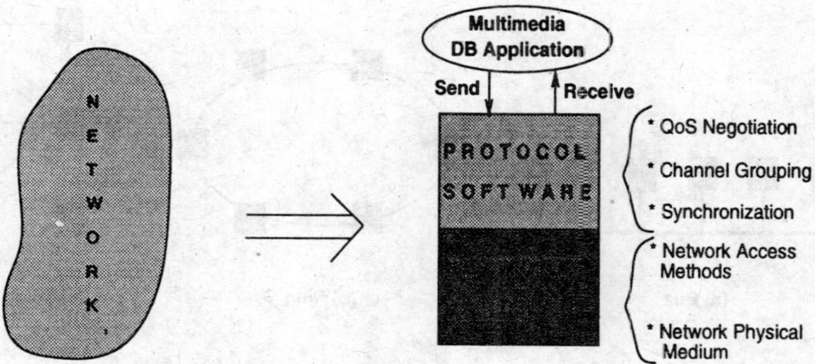

Figure 7.12 Network Support for Multimedia Communication

- **Network Hardware :** The components of the hardware, physical medium and the corresponding access method used, influence the QoS that can be offered to multimedia applications.
- **Network Protocol Software :** support has to be provided for QoS negotiation, group of channels, and synchronization of object transfer.

7.3.1 Network Hardware Considerations

Better network hardware platform is needed in order to offer high bandwidth. The network hardware provides access to the physical network medium. The physical network medium can be optic fibre, copper wire, etc. The type of the physical medium determines the maximum possible amount of information that can be carried in a unit time by a network. For example, optic fibre networks operate in the range of Gigabits per second while copper wire networks operate in the range of Megabits per second. The physical medium is shared by the computers connected to the network. Hence, the access to the physical medium for a computer (server or client) has to be regulated in order to ensure fairness. This access regulation is termed as the *Medium Access Protocol.* The following characteristics of the network medium and the access methods have a distinct bearing on the offered QoS parameters :

- Network topology
- Network bandwidth
- Network access control mechanism

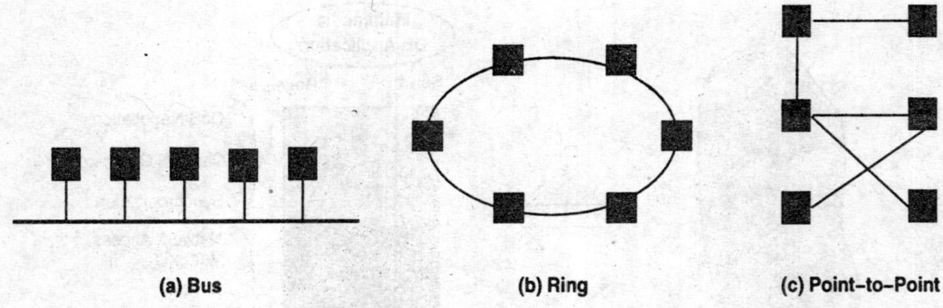

Figure 7.13 Popular Network Topologies

- Priority control schemes for network access

Throughput guarantees by the network service provider depend on the bandwidth offered by the network hardware as well as the regulated access to the physical medium. The network bandwidth typically ranges from medium speed (operating at a few Mbits/s) to very high speeds (operating at several hundreds of Mbits/s). The delay guarantees depend on the medium access control methods and the availability of priority control schemes to access the physical medium. Now, we discuss some popular network topologies and network access protocols.

Network Topology

Network topology refers to the way in which the computers are connected to form the network. The popular network topologies are bus and ring topologies. Figure 7.13 (a) and (b) show possible bus and ring topologies. Ethernet and Token Bus networks use bus topology while Token Ring and FDDI (Fiber Distributed Data Interface) use ring topology. Figure 7.13 (c) shows a point-to-point network topology. This network employs switches to transfer data from one node to another. ATM (Asynchronous Transfer Mode) networks use this point-to-point network topology.

Network Access Protocols

The network topology influences the type of access to the network medium. For example, in the point-to-point networks, the network medium is shared by the computers *only* through the switching nodes. Hence, the access to the network is regulated by the switching nodes. In the bus and ring topologies, the medium is directly shared by the computers and hence the access control strategies are different.

Bus and Ring Topologies : The commonly used access control protocols for the bus and ring topologies are :

- Random access control
- Token based access control

The random access control method is used by networks such as Ethernet. The strategy for random access is called *Listen While Talking.* Here, computers connected to the network are allowed to communicate data *whenever* (i.e., random) they need to. The computers also listen to the network while they are communicating. This strategy of random access results in collisions of information when *multiple* computers *decide* to communicate data *simultaneously.* Since the computers also listen while they are communicating, they can detect the occurrence of collisions. When such collisions are detected, the computers stop the communication. They wait for a random period of time before retrying the communication. However, collisions can possibly occur when the computers retry to communicate the information. Due to the possibility of repeated collisions, it is difficult to guarantee delivery of information. Hence, it is difficult to guarantee QoS parameters such as throughput and delay to multimedia database applications, using random medium access control.

Token based access methods are used by token ring and token bus networks. Here, a permit to communicate information, in the form of a token, circulates in the network. A computer is allowed to communicate information *only when* it has the token. The communication of information is done only for an allotted time period. After the allotted time period, the token is passed to the *next* computer. The token based access methods provide a regulated access to the network medium and hence it is possible to guarantee QoS parameters, such as throughput and delay to multimedia database applications. Token based access control is used by FDDI networks. Priority schemes for circulating the tokens can also be employed to provide better control to network medium. Systems with higher priorities may be allowed to transmit data more often than others. This facility can help in transmitting data (such as video) in such a way that its real-time requirements are met.

Point-to-Point Topology : ATM is an access control method that is being popularly used for point-to-point networks. Here, switches multiplex the information to various output channels. ATM uses Asynchronous Time Division multiplexing. The unit of multiplexing is called an ATM cell, a small packet of constant length, 53 bytes. Communication channel between a server and a client is established by concatenating several point-to-point links. An ATM Adaptation Layer (AAL) has been defined to enhance the services provided by the ATM layer. The AAL helps in providing different modes of service to the applications. It can provide QoS guarantees such as throughput and delay to multimedia database applications.

7.3.2 Network Protocol Support

The pre-defined format in which the communication exchange between a server and a client takes place is called the network protocol. Standard network protocols such as TCP/IP and ISO Open Systems Interconnection (OSI), as well as proprietary protocols such as DECnet are commonly used for communication between servers and clients. However, network protocol support for multimedia applications are different from those needed for traditional applications, such as file transfer, electronic mail, and remote login. As discussed earlier, network protocol support for multimedia applications is provided in terms of :

- Channel group services
- QoS Negotiation
- Synchronization of object transfer

New protocols or enhancements to existing protocols have been suggested in order to provide communication support for multimedia database applications. Here, we shall discuss in brief the features of the following protocols:

Orchestration Services in ISO-OSI : Orchestration services in ISO-OSI aim at providing channel grouping facility in the existing OSI model (refer [84, 85]). The orchestrator provides services for starting and stopping group of communication channels as well as services for controlling and monitoring individual channel. Controlling a group of channels includes services for dynamically joining/leaving the group and for regulating the QoS needs of individual channels. These orchestration services include provision for specification and negotiation of QoS parameters as well as their tolerance levels during channel establishment.

ST-II & RSVP : In order to provide services such as QoS guarantees and synchronization of object transfers, network service provider has to reserve buffer resources for the communication channel to be established. Internet Stream protocol version II (ST-II) and ReSerVation Protocol (RSVP) address this issue of resource reservation in internetworks. Interested readers can refer [39, 104] for further details.

7.4 CONCLUDING REMARKS

Communication channels need to be established between server(s) and a client for communication of multimedia information. Multimedia information is generally voluminous in nature, with temporal constraints for the presentation of information. This results in the need for guarantees from the network service provider for Quality of Service (QoS) to the multimedia applications. These QoS parameters are specified in terms of parameters such as throughput, delay, delay-jitter, and packet-loss probabilities. The characteristics of the physical medium influence the maximum throughput that can be supported by the network. The access method employed specifies the mechanism of sharing the maximum possible network bandwidth among the computers connected to the network.

The network service provider interfaces with the multimedia database applications in order to facilitate access to distributed information. A client needs to specify and negotiate its QoS requirements with the network service provider. Multimedia database applications also may need channel group services to communicate individual media information from server(s) to client. Transfer of information such as audio and video might have to be synchronized and this synchronization might have to be enforced between channels carrying the information. The services that are required from network hardware and protocol software for a multimedia database application are summarized in Table 7.1.

Figure 7.14 shows a simple block diagram of a communication manager. The connection establishment module, taking the QoS specification as input, establishes communication channel(s) between the client and the server. It uses the services provided by the network access protocol in order to carry out the

	Technique	Features
Physical Medium		
	Coaxial cables	Order of Mega bits per second
	Fiber optic cables	Order of Giga bits per second
Network Access Method	Random access control	Cannot be used for guaranteeing QoS parameters
	Token based access	Can guarantee QoS parameters
	Asynchronous Transfer	— do —
Network Protocol Services		(i) QoS Negotiation (ii) Channel Group Services (iii) Synchronization of Object Transfer Across Communication Channels

Table 7.1 Network Support For Multimedia Databases

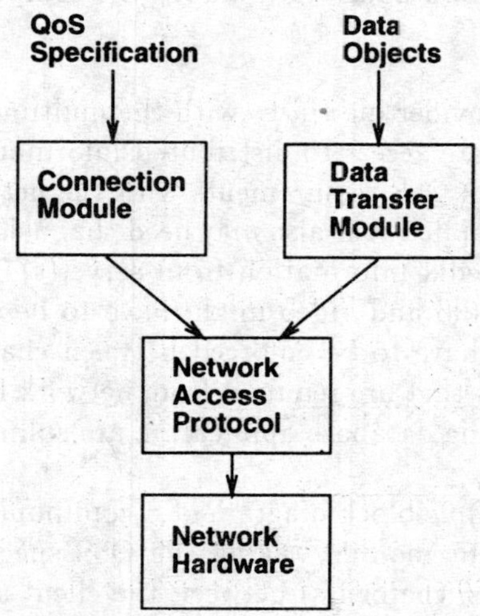

Figure 7.14 Components of Communication Manager

job. The data transfer module then helps in communicating the multimedia information.

Bibliographic Notes

Retrieval schedule generations for multimedia database applications are discussed in [56, 58, 77, 164]. [56, 58, 77] discuss derivation of schedules based on Petri nets specifications of the temporal characteristics. [164] presents techniques for deriving flexible retrieval schedules based on difference constraints specification of temporal characteristics.

Quality of Service (QoS) requirements of communication channels are described in [36, 41, 76, 78, 82]. QoS negotiation issues are presented in [112, 159, 165].

Orchestration services in ISO-OSI are introduced in [84, 85]. Internet Stream protocol version II (ST-II) and ReSerVation Protocol (RSVP) address the issue of resource reservation in internetworks [39, 104]. Requirements of a group of communication channels are discussed in [106]. Synchronization protocols for multimedia database applications are presented in [56, 58]. In [165], Multimedia Application Protocol (MMAP) has been proposed for handling user interactions and for interacting with the network service provider.

8

MMDBMS ARCHITECTURE

In the previous chapters, we discussed the issues and the techniques used in building multimedia database management systems (MMDBMS). In this chapter, we summarize by providing a simple architecture of a distributed MMDBMS that uses the various components discussed so far.

8.1 DISTRIBUTED MMDBMS ARCHITECTURE

Figure 8.1 shows the architecture of a typical distributed multimedia database management system. This architecture comprises of the various components of the MMDBMS discussed in the previous chapters. The architecture shows a multimedia database server and a client connected by a computer network.

8.1.1 MMDBMS Server Components

A typical multimedia database management system has the following components :

- **Storage Manager :** handles the storage and retrieval of different media objects composing a database (described in Chapter 2, Section 2.5). The metadata as well as the index information related to media objects are also handled by this module.

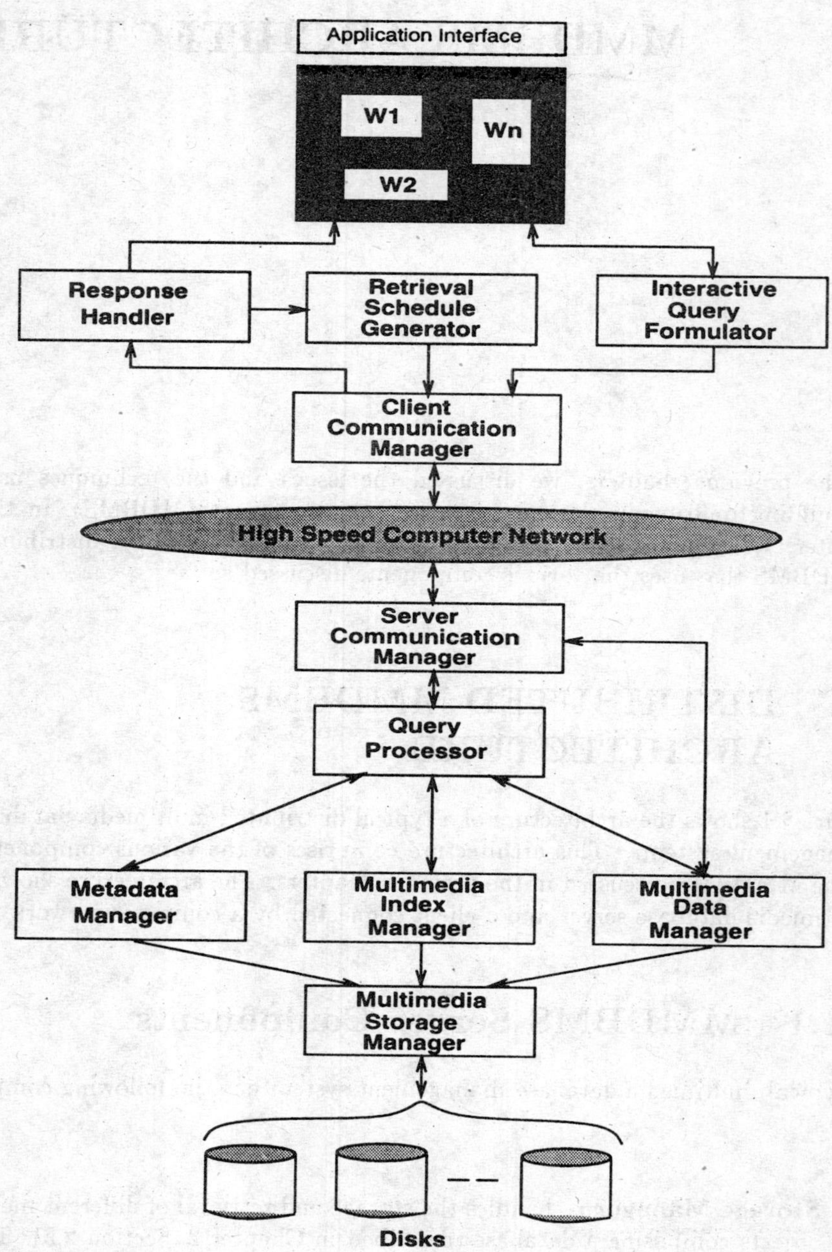

Figure 8.1 Architecture of a Typical Distributed Multimedia Database Management System

- **Metadata Manager :** deals with creating and updating metadata associated with multimedia objects (described in Chapter 3, Section 3.6). The metadata manager provides relevant information to the query processor in order to process a query.

- **Index Manager :** supports formulation and maintenance of faster access structures for multimedia information (described in Chapter 4, Section 4.5).

- **Data Manager :** helps in creation and modification of multimedia objects (described in Chapter 5, Section 5.5). It also helps in handling temporal and spatial characteristics of media objects. Metadata manager, index manager, and object manager access the required information through the storage manager.

- **Query Processor :** receives and processes user queries (described in Chapter 6, Section 6.3). This module utilizes information provided by metadata manager, index manager, and data manager for processing a query. In the case where objects are to be presented to the user as part of the query response, the query processor identifies the sequence of the media objects to be presented. This sequence along with the temporal information for object presentation is passed back to the user.

- **Communication Manager :** handles the interface to the computer network for server-client interaction (described in Chapter 7, Section 7.4). This module, depending on the services offered by the network service provider, reserves the necessary bandwidth for server-client communication, and transfers queries and responses between the server and the client.

8.1.2 MMDBMS Client Components

A typical client in the multimedia database management system has the following components :

- **Communication Manager :** manages the communication requirements of a multimedia database client. Its functionalities are the same as that of the server communication manager.

- **Retrieval Schedule Generator :** determines the schedule for retrieving media objects (described in Chapter 7, Section 7.1.3). For the retrieval schedule generator, the response handler provides information regarding the objects composing the response and the associated temporal information. The retrieval schedule generator, based on the available buffer and network throughput, determines the schedule for object retrieval. This retrieval schedule is used by the communication manager to download media objects in the specified sequence.

- **Response Handler :** interacts with the client communication manager in order to identify the type of response generated by the server. If the response comprises of the information regarding the objects to be presented, then the information is passed to the retrieval schedule generator for determining the retrieval sequence. If the response comprises of the information regarding modifying the posed query or a null answer, the response is passed to the user.
- **Interactive Query Formulator :** helps a user to frame an appropriate query to be communicated to a database server (as discussed in Chapter 6, Section 6.3). This module takes the response from the server (through the response handler module) in order to reformulate the query, if necessary.

8.2 IMPLEMENTATION CONSIDERATIONS

The implementation of different modules composing a distributed MMDBMS depends on the hardware resources, operating systems, and the services offered by computer networks.

Hardware resources : The available hardware resources influence both client and server design. On the server side, the available disk space limits the size of a multimedia database that can be handled. The number of queries that can be handled and the query response time depends, to a certain extent, on the speed of the system. On the client side, buffer availability for media objects retrieval is influenced by available hardware resources. User interface for multimedia presentation is also influenced by hardware characteristics such as monitor resolution, width and height.

Operating System : Multimedia databases are composed of continuous media objects. The retrieval and presentation of these media objects have an associated time constraint. Hence, real-time features might be needed as part of the operating system. Also, the file system needs to handle large media objects, as discussed in Chapter 2.

Computer Network : Services offered by computer networks influences the retrieval of media objects by a client system. If guaranteed throughput is not offered by the network service provider, the client may not be able to retrieve objects at the required time. This may lead to an object presentation sequence that differs from the temporal requirements specified in the database. Also, the buffer requirements in the client system depend on the throughput offered by the network service provider, as discussed in Chapter 7.

8.3 CONCLUDING REMARKS

Multimedia database management systems are experiencing rapid growth and development. This field spans over a wide area of research interests : real-time operating systems, computer networks, signal processing (audio, image, and video), and the database aspects such as index generation and query processing.

Real-time process scheduling techniques help in guaranteeing the temporal requirements. Very large file systems have been made possible due to the development of large disk arrays. Different disk scheduling techniques are being developed to handle guaranteed retrieval of multimedia information. In the area of computer networks, large network bandwidths have become realities. Network hardware such as ATM are being developed to provide guaranteed and fast access to remote information. Network protocol software developments are being done to handle resource reservation, multicasting, and security considerations.

Developments in the area of signal processing are contributing towards better handling of media objects such as audio, image, and video. Since the information in these media objects are inherently binary in nature, signal processing techniques help in automatic/semi-automatic generation of metadata associated with these objects. A good amount of work has been done in this area but a lot more needs to be done in order to achieve the dream of a computer understanding multimedia objects like human beings.

Database aspects that require further attention for handling multimedia information include index structures and query processing. Multiple metadata features of a media object have to be appropriately indexed. Mapping functions and spatial index structures have been developed to handle these issues. Query processing for multimedia databases involves handling query-by-example and partial matching responses. Processing queries-by-example involves both signal processing and efficient query processing techniques, for handling the various possible media objects. Handling partial matching responses implies the ability to select only those responses which are similar.

Researchers in the above areas are actively contributing new concepts and techniques, leading to an ever-changing multimedia database environment.

REFERENCES

[1] M.E. Maron and J.L. Kuhns, "On Relevance, Probabilistic Indexing, and Information Retrieval", Journal of ACM, Vol. 7, 1960, pp. 216-244.

[2] A.V. Aho and M.J. Corasick, "Fast Pattern Matching : An Aid to Bibliographic Search", Communications of ACM, Vol. 18, No. 6, June, 1975, pp. 333-340.

[3] C.T. Yu and G. Salton, "Precision Weighting : An Effective Automatic Indexing Method", Journal of ACM, Vol. 23, 1976, pp. 76-88.

[4] D.E. Knuth, J.H. Morris and V.R. Pratt, "Fast Pattern Matching in Strings", SIAM Journal of Computer, Vol. 6, No. 2, June 1977, pp. 323-350.

[5] R.S. Boyer and J.S. Moore, "A Fast String Searching Algorithm", Communications of ACM, Vol. 20, No. 10, October 1977, pp. 762-772.

[6] H. Tamura, S. Morai and T. Yamawaki, "Texture Features Corresponding to Visual Perception", IEEE Transactions on Systems, Man, and Cybernatics, SMC-8(6), pp. 460-473, 1978.

[7] K.R. Castleman, *Digital Image Processing*, Prentice-Hall Inc., Englewood Cliffs, NJ, 1979.

[8] J.L. Peterson, *Petri Net Theory and The Modeling of Systems*, Prentice-Hall Inc., 1981.

[9] F.R. Chen and M.M. Withgott, "The Use of Emphasis to Automatically Summarize A Spoken Discourse", Proc. International Conference on Acoustics, Speech and Signal Processing, San Francisco, California, March 1982.

[10] W. Reisig, *Petri Nets : An Introduction*, Springer-Verlag Publication, 1982.

[11] K.S. Fu, *Syntactic Pattern Recognition and Applications*, Prentice-Hall Inc., Englewood Cliffs, New Jersy, 1982.

[12] J.E. Coolahan. Jr., and N. Roussopoulos, "Timing requirements for Time-Driven Systems Using Augmented Petri Nets", IEEE Trans. Software Eng., Vol. SE-9, Sept. 1983, pp 603-616

[13] J.F. Allen, "Maintaining Knowledge about Temporal Intervals", Communications of the ACM, November 1983, Vol. 26, No. 11, pp. 832-843.

[14] H. Samet, "The Quadtree and Related Hierarchical Data Structures", Computing Surveys, Vol. 16, No. 2, June 1984, pp. 187-260.

[15] M. Chock, A.F. Cardenas and A. Kingler, "Data Structure and Manipulation Capabilities of a Picture Database Management System", IEEE Transactions on Pattern Analysis and Machine Intelligence, 6(4), pp. 484-492, 1984.

[16] A. Guttman, "R-Trees : A Dynamic Index Structure for Spatial Searching", ACM SIGMOD, Intl. Conference on the Management of Data, 1984, pp. 47-57.

[17] ISO, "PHIGS - Programmers Hierarchical Interface to Graphics Systems", ISO/TC97/SC5/WG2/N305, 1984.

[18] C. Faloutsos, "Access Methods for Text", ACM Computing Surveys, Vol. 17, No. 1, March 1985.

[19] W.M. Zuberek, "M-Timed Petri nets, Priorities, Pre-Emptions and Performance Evaluation of Systems", Advances in Petri nets 1985, Lecture Notes in Computer Science (LNCS 222), Springer-Verlag, 1985.

[20] T. Sellis, N. Roussopoulos and C. Faloutsos, "The R+-Tree : A Dynamic Index For Multi-dimensional Objects", Proc. 13th VLDB Conference, Brighton, U.K., 1987, pp. 507-518.

[21] F. Preparata and M. Shamos, *Computational Geometry : An Introduction*, Springer-Verlag, NY, 1985.

[22] S. Christodoulakis, M. Theodoridou, F. Ho, M. Papa and A. Pathria, "Multimedia Document Presentation, Information Extraction, and Document Formation in MINOS : a Model and a System", ACM Transactions on Office Information Systems, 4(4), pp. 345-383, October 1986.

[23] ISO, "Information Processing - Text and Office Systems - Standardized Generalized Markup Language (SGML)", International Standards Organization, ISO 8879-1986(E) edition, 1986.

[24] S. Gibbs, D. Tsichritzis, A. Fitas, D. Konstantas and Y. Yeorgaroudakis, "MUSE : A Multimedia Filing System", IEEE Scftware, March 1987.

[25] S.K. Chang, Q.Y. Shi and C.W. Yan, "Iconic Indexing by 2D Strings", IEEE Trans. Pattern Analysis, Machine Intelligence, Vol. 9, 1987, pp. 413-428.

[26] R. Gonsalez and P. Wintz, *Digital Image Processing*, Addison-Wesley, Reading, MA, 1987.

[27] D.A. Patterson, G. Gibson and R.H. Katz, "A Case for Redundant Arrays of Inexpensive Disks (RAID)", Proceedings of the ACM Conference on Management of Data, pp. 109 - 116, June 1988.

[28] E. Bertino, F. Rabiti and S. Gibbs, "Query Processing in a Multimedia Document System", ACM Transactions on Office and Information Systems, 6, pp. 1-41, 1988.

[29] P.D. Stotts and R. Frutta, "Petri-Net-Based Hypertext : Document Structure With Browsing Semantics", ACM Trans. on Office Information Systems, Vol. 7, No. 1, Jan 1989, pp. 3-29.

[30] T.J. Lehman and B.G. Lindsay, "The Starburst Long Field Manager", Proceedings of Fifteenth International Conference on Very Large Databases, Amsterdam, 1989, pp. 375-383.

[31] M. Ioka, "A Method for Defining the Similarity of Images on the Basis of Color Information", Technical Report RT-0030, IBM Tokyo Research Lab, Japan, 1989.

[32] H. Sameet, *The Design and Analysis of Spatial Data Structures*, Addison-Wesley, 1989.

[33] L.R. Rabiner, "A Tutorial on Hidden Markov Models and Selected Applications", Proc. IEEE, Vol. 77, No. 2, February 1989.

[34] T.D.C. Little and A Ghafoor, "Synchronization and Storage Models for Multimedia Objects", IEEE Journal on Selected Areas of Communication, Vol. 8, No. 3, April 1990, pp. 413-427.

[35] R. Steinmetz, "Synchronization Properties in Multimedia Systems", IEEE J. on Selected Areas of Communication, Vol. 8, No. 3, April 1990, pp. 401-412.

[36] D. Ferrari and D.C. Verma, "A Scheme for Real-Time Channel Establishment in Wide Area Networks", IEEE Journal on Selected Areas in Communication, Vol. 8, No. 3, April, 1990, pp. 368-379.

[37] R.D. Peacocke and D.H. Graf, "An Introduction to Speech and Speaker Recognition", IEEE Computer, August 1990, pp. 26-33.

[38] P.D. Stotts and R. Frutta, "Temporal Hyperprogramming", Journal of Visual Languages and Computing, Sept. 1990, pp. 237-253.

[39] C. Topolcic, "Experimental Internet Stream Protocol : Version 2 (ST-II)", Internet RFC 1190, October, 1990.

[40] Suh-Yin Lee and Fang-Jung Hsu, "2D C-String : A New Spatial Knowledge Representation For Image Database System", Pattern Recognition, 23(10), pp. 1077-1087, 1990.

[41] D. Ferrari, "Client Requirements For Real-Time Communication Services", IEEE Communication Magazine, Vol. 28, No. 11, November, 1990, pp. 65-72.

[42] E. Bertino and F. Rabitti, "The MULTOS Query Language", In Multimedia Office Filing : The MULTOS Approach, pp. 53-74, Elsevier, 1990.

[43] E. Bertino and F. Rabitti, "Query Processing in MULTOS", In Multimedia Office Filing : The MULTOS Approach, pp. 273-295, Elsevier, 1990.

[44] F. Rabbitti and P. Savino, "Retrieval of Multimedia Documents by Imprecise Query Specification", Proc. of the Int. Conference on Extended Database Technologies, pp. 203-218, 1990.

[45] D. Cutting and J. Pedersen, "Optimizations for Dynamic Inverted Index Maintenance", Proceedings of ACM SIGIR, 1990, pp. 405-411.

[46] N. Beckmann, H.P. Kriegel, R. Schneider and B. Seeger, "The R^*-tree: An Efficient and Robust Access Method for Points and Rectangles", ACM SIGMOD, May 1990, pp. 322-331.

[47] M.J. Swain and D.H. Ballard, "Color Indexing", International Journal of Computer Vision, Vol. 7, No. 1, 1991, pp. 11-32.

[48] S. Gibbs, "Composite Multimedia and Active Objects", Proc. OOPSLA '91, pp. 97-112.

[49] G.K. Wallace, "The JPEG Still Picture Compression Standard", Communications of the ACM, Vol. 34, No. 4, April 1991, pp. 30-44.

[50] D. Le Gall, "MPEG : A Video Compression Standard for Multimedia Applications", CACM, 34(4):46-58, April 1991.

[51] H.V. Jagdish, "A Retrieval Technique For Similar Shapes", International Conference on Management of Data, SIGMOD '91, pp. 208-217, May 1991.

[52] H. Turtle and W.B. Croft, "Evaluation of an Inference Network-Based Retrieval Model", ACM Transactions on Information Systems, Vol. 9, No. 3, July 1991, pp. 187-222.

[53] N. Fuhr and C. Buckley, "A Probabilistic Learning Approach for Document Indexing", ACM Transactions on Information Systems, Vol. 9, No. 3, July 1991, pp. 223-248.

[54] S. Gauch and J.B. Smith, "Search Improvement via Automatic Query Reformulation", ACM Transactions on Information Systems, Vol. 9, No. 3, July 1991, pp. 249-280.

[55] E.A. Fox, Q.F. Chan, A.M. Daoud and L.S. Heath, "Order-Preserving Minimal Perfect Hash Functions and Information Retrieval", ACM Transactions on Information Systems, Vol. 9, No. 3, July 1991, pp. 281-308.

[56] T.D.C. Little, *Synchronization For Distributed Multimedia Database Systems*, PhD Dissertation, Syracuse University, August 1991.

[57] T.D.C. Little, A. Ghafoor, C.Y.R. Yen, C.S. Chen and P.B. Berra, "Multimedia Synchronization", IEEE Data Engineering Bulletin, Vol. 14, No. 3, September 1991, pp. 26-35.

[58] T.D.C. Little and A. Ghafoor, "Multimedia Synchronization Protocols for Broadband Integrated Services", IEEE Journal on Selected Areas in Communication, Vol. 6, No. 9, December, 1991, pp. 1368-1382.

[59] C. Chang and S.Y. Lee, "Retrieval of Similar Pictures on Pictorial Databases", Pattern Recognition, 24(7), pp. 675-680,1991.

[60] L.F. Cabrera and D.D.E. Long, "Swift : Using Distributed Disk String to Provide High I/O Data Rates", Computer Systems, 4(4), pp. 405 - 436, Fall 1991.

[61] C.P. Kolovson and M. Stonebraker, "Segment Indexes : Dynamic Indexing Techniques for Multi-Dimensional Interval Data", ACM SIGMOD, May 1991, pp. 138-147.

[62] P. Venkat Rangan and H.M. Vin, "Multimedia Conferencing as a Universal Paradigm For Collaboration", In Multimedia Systems, Applications and Interactions, (Chapter 14), Lars Kjelldahl (editor), Springer-Verlag, Germany, 1991.

[63] D.S. Bloomberg, "Multiresolution Morphological Approach to Document Image Analysis", Proceedings of the International Conference on Document Analysis and Recognition, Saint-Malo, France, September 1991.

[64] H.M. Vin et al, "Multimedia Conferencing in the Etherphone Environment", IEEE Computer, Special Issue on Multimedia Information Systems, Vol. 24, No. 10, October 1991, pp. 69-79.

[65] A. Nagasaka and Y. Tanaka, "Automatic Video Indexing and Full-motion Search for Object Appearnces", Proceedings of 2nd Working Conference on Visual Databases, Budapest, October 1991, pp. 119-133.

[66] H.M. Vin, P. Venkat Rangan and S. Ramanathan, "Hierarchical Conferencing Architectures for Inter-Group Multimedia Collaboration", Proc. of the ACM Conf on Organizational Computing Systems (COCS'91), Atlanta, Georgia, November 1991.

[67] S.R. Newcomb, N.A. Kipp and V.T. Newcomb, "The HyTime : Hypermedia/Time-based Document Structuring Language", Communications of the ACM, Vol. 34, No. 11, 1991.

[68] P. Venkat Rangan and D.C. Swinehart, "Software Architecture for Integration of Video Services in the Etherphone System", IEEE Journal on Selected Areas in Communications, Vol. 9, No. 9, December 1991.

[69] L.D. Wilcox and M.A. Bush, "HMM-Based Wordspotting for Voice Editing and Indexing", Proceedings of European Conference on Speech Communication and Technology, 1991, pp. 25-28.

[70] ISO, "Document Style Semantics and Specification Language (DSSSL)", International Standards Organization, ISO/IEC 10179-1991 (E) edition, 1991.

[71] J.L. Mitchell and W.B Pennebaker, "Evolving JPEG Color Data Compression Standards", M. Nier, M.E. Courtot (eds.): Standards for Electronic Imaging Systems, SPIE, Vol. CR37, 1991, pp. 68-97.

[72] ISO, "Information Technology - Hypermedia/Time-based Structuring Language (HyTime)", International Standards Organization, ISO/IEC 10744-1992 (E) edition, 1992.

[73] A. samal and P.A. Iyengar, "Automatic Recognition and Analysis of Human Faces and Facial Expressions : A Survey", Pattern Recognition, Vol. 25, Jan. 1992, pp. 65-77.

[74] J. Stefani, L. Hazard and F. Horn, "Computaticnal Model for Distributed Multimedia Applications Based On a Synchronous Programming Language", Computer Communication, Butterworth-Hienmann Ltd., Vol. 15, No. 2, March 1992, pp.114-128.

[75] S.Y. Lee and F.J. Hsu, "Spatial Reasoning and Similarity Retrieval of Images Using 2D C-String Knowledge Representation", Pattern recognition, Vol. 25, No. 3, 1992, pp. 305-318.

[76] D. Ferrari, "Delay Jitter Control Scheme For Packet-Switching Internetworks", Computer Communications, Vol. 15, No. 6, July/August, 1992, pp. 367-373.

[77] T.D.C. Little and A. Ghafoor, "Scheduling of Bandwidth-Constrained Multimedia Traffic", Computer Communication, Butterworth-Heinemann, July/August 1992, pp. 381-388.

[78] ISO/IEC/JTC 1/SC21/WG 1N 1201, "A Suggested QoS Architecture For Multimedia Communications", September, 1992.

[79] W.I. Grosky, P. Neo and R. Mehrotra, "A Pictorial Index Mechanism For Model-Based Matching", Data and Knowledge Engineering, 8, pp. 309-327, 1992.

[80] I. Kamel and C. Faloutsos, "Parallel R-trees", Proceedings of ACM SIGMOD '92, 1992, pp. 195-204.

[81] M. Moran and R. Gusella, "System Support For Efficient Dynamically Configurable Multi-Party Interactive Multimedia Applications", Proc. of 3rd Intl. Workshop on Network, Operating System Support for Digital Audio and Video, San Diego, California, November 1992.

[82] ISO/IEC/JTC 1/SC21/WG N 7430, "Working Draft of the Technical Report on Multimedia and Hypermedia : Model and Framework", November, 1992,

[83] U. Glavitsch and P. Schäuble, "A System for Retrieving Speech Documents", Proc. of ACM SIGIR Conference on R&D in Information, Denmark, 1992, pp. 168-176.

[84] A. Campbell, G. Coulson, F. Garcia and D. Hutchison, "A Continuous Media Transport and Orchestration Service", Proc. of ACM SIGCOMM'92, 1992, pp. 99-110.

[85] A. Campbell *et al*, "Orchestration Services For Distributed Multimedia Synchronization", Proc. 4th IFIP Conference On High Performance Networking", Liege, Belgium, December, 1992.

[86] M. Caudill and C. Butler, *Understanding Neural Networks : Computer Explorations Vol. 1 and 2*, MIT Press, Cambridge, 1992.

[87] M.A. Hearst, "TextTiling : A Quantitative Approach to Discourse Segmentation", Technical Report 93/24, University of California, Berkeley, 1993.

[88] M.A. Hearst and C. Plaunt, "Subtopic Structuring for Full-Length Document Access", Proceedings of ACM SIGIR, Pittsburgh, 1993, pp. 59-68.

[89] P. Schäuble, "SPIDER : A Multiuser Information Retrieval System for Semistructured and Dynamic Data", Proceedings of ACM SIGIR, 1993, pp. 318-327.

[90] H.M. Vin and P. Venkat Rangan, "Designing a Multi-User HDTV Storage Server", IEEE Journal on Selected Areas on Communication, January 1993.

[91] H.J. Zhang, A. Kankanhalli and S.W. Smoliar, "Automatic Partitioning of Video", Multimedia Systems, 1(1), 1993, pp. 10-28.

[92] M.C. Buchanan and P.T. Zellweger, "Automatically Generating Consistent Schedules for Multimedia Documents", ACM/Springer-Verlag Journal of Multimedia Systems, Vol. 1, No. 2, 1993.

[93] R. Price, "MHEG : An Introduction to the Future International Standard for Hypermedia Object Interchange", ACM Multimedia '93, pp. 121 - 128, 1993.

[94] H. Ishikawa *et al*, "The Model, Language, and Implementation of an Object-Oriented Multimedia Knowledge Base Management System", ACM Transactions on Database Systems, Vol. 18, No. 1, March 1993, pp. 1-50.

[95] S. Kuo and O.E. Agazzi, "Machine Vision for Keyword Spotting Using Pseudo 2D Hidden Markov Models", Proceedings of International Conference on Acoustics, Speech and Signal Processing, Minneapolis, Minnesota, April 1993.

[96] F. Arman, A. Hsu and M. Chiu, "Image Processing on Compressed Data for Large Video Databases", Proceedings of First ACM Conference on Multimedia 1993, Anaheim, California, pp. 267-272.

[97] H.J. Zhang, A. Kankanhalli and S.W. Smoliar, "Automatic Partitioning of Full-Motion Video", Multimedia Systems, Springer-Verlag, 1(1), 1993, pp. 10-28.

[98] T.D.C. Little and A. Ghafoor, "Interval-Based Conceptual Models for Time-Dependent Multimedia Data", IEEE Transactions on Knowledge and Data Engineering, Vol. 5, No. 4, August 1993, pp. 551-563.

[99] P. Venkat Rangan and H.M. Vin, "Efficient Storage Techniques for Digital Continuous Media", IEEE Transactions on Knowledge and Data Engineering, Vol. 5, No. 4, August 1993, pp. 564-573.

[100] J. Riedl, V. Mashayekhi, J. Schnepf, M. Claypool and D. Frankowski, "SuiteSound : A System for Distributed Collaborative Multimedia", IEEE Transactions on Knowledge and Data Engineering, Vol. 5, No. 4, August 1993, pp. 600-609.

[101] J.R. Bach, S. Paul and R. Jain, "A Visual Information Management System for the Interactive Retrieval of Faces", IEEE Transactions on Knowledge and Data Engineering, Vol. 5, No. 4, August 1993, pp. 619-628.

[102] E. Oomoto and K. Tanaka, "OVID : Design and Implementation of a Video-Object Database System", IEEE Transactions on Knowledge and Data Engineering, Vol. 5, No. 4, pp. 629-641, 1993.

[103] A.F. Cardenas, I.T. Ieong, R.K. Taira, R. Barker, C.M. Breant, "The Knowledge-Based Object-Oriented PICQUERY+ Language", IEEE Transactions on Knowledge and Data Engineering, 5(4), August 1993, pp. 644-657.

[104] L. Zhang *et al*, "RSVP: A New Resource ReSerVation Protocol", IEEE Network Magazine, September, 1993.

[105] ISO IEC JTC 1: Information Technology - Coding of Moving Pictures and Associated Audio for Digital Storage Media up to about 1.5 Mbits/s; International Standard ISO/IEC IS 11172, 1993.

[106] A. Gupta and M. Moran, "Channel Groups - A Unifying Abstraction for Specifying Inter-stream Relationships", ICSI TR-93-015, Technical Report, Berkeley, 1993.

[107] W.I. Gorsky, "Multimedia Information Systems : A Tutorial", IEEE Multimedia, Vol. 1, No. 1, 1994.

[108] E. Binaghi, I. Gagliardi, R. Schettini, "Image Retrieval Using Fuzzy Evaluation of Color Similarity", International Journal of Pattern Recognition and Artificial Intelligence, 8(4), 1994, pp. 945-968.

[109] L. Li, A. Karmouch and N.D. Georganas, "Multimedia Teleorchestra With Independent Sources : Part 1 - Temporal Modeling of Collaborative Multimedia Scenarios", ACM/Springer-Verlag Journal of Multimedia Systems, Vol. 1, No. 4, February 1994, pp.143-153.

[110] L. Li, A. Karmouch and N.D. Georganas, "Multimedia Teleorchestra With Independent Sources : Part 2 - Synchronization Algorithms", ACM/Springer-Verlag Journal of Multimedia Systems, Vol. 1, No. 4, February 1994, pp.153-165.

[111] S.W. Smoliar and H.J. Zhang, "Content-Based Video Indexing and Retrieval", IEEE Multimedia, 1(2), 1994, pp. 62-72.

[112] S.V. Raghavan, B. Prabhakaran and S.K. Tripathi, "Quality of Service Negotiation For Orchestrated Multimedia Presentation", Proceedings of High Performance Networking Conference HPN 94, Grenoble, France, June 1994, pp.217-238. Also available as Technical Report CS-TR-3167, UMIACS-TR-93-113, University of Maryland, College Park, USA, October 1993.

[113] B. Prabhakaran and S.V. Raghavan, "Synchronization Models For Multimedia Presentation With User Participation", ACM/Springer-Verlag Journal of Multimedia Systems, Vol.2, No. 2, August 1994, pp. 53-62. Also in the Proceedings of the First ACM Conference on Multimedia Systems, Anaheim, California, August 1993, pp.157-166.

[114] B. Furht, "Multimedia Systems : An Overview", IEEE Multimedia, Vol. 1, No. 1, Spring 1994, pp. 47-59.

[115] T.S. Chua, S.K. Lim, H.K. Pung, "Content-based Retrieval of Segmented Images", Proceedings of ACM Multimedia'94, 1994.

[116] C. Faloutsos, R. Barber, M. Flickner, J. Hafner, W. Niblack, D. Petkovic and W. Equitz, "Efficient and Effective Querying by Image Content", Journal of Intelligent Information Systems, 3(3/4), pp. 231-262, July 1994.

[117] N. Dimitrova and F. Golshani, "RX for Semantics Video Database Retrieval", Proceedings of ACM Multimedia'94.

[118] S. Berson, S. Ghandeharizadeh, R. Muntz and X. Ju, "Staggered Striping in Multimedia Information Systems", Proceedings of ACM SIGMOD '94, Minneapolis, 1994, pp. 79-90.

[119] A. Tomasic, H.G. Molina and K. Shoens, "Incremental Updates of Inverted Lists for Text Document Retrieval", Proceedings of ACM SIGMOD'94, 1994, pp. 289-300.

[120] A. Laursen, J. Olkin and M. Porter, "Oracle Media Server : Providing Consumer Based Interactive Access to Multimedia Data", Proceedings of ACM SIGMOD '94, Minneapolis, 1994, pp. 470-477.

[121] W. Klas and A. Sheth, editors, "Special Issue on Meta-data for Digital Media", No. 4, ACM SIGMOD RECORD, December 1994.

[122] K. Bohms and T.C. Rakow, "Metadata for Multimedia Documents", No. 4, ACM SIGMOD RECORD, December 1994, pp. 21-26.

[123] R. Jain and A. Hampapur, "Metadata in Video Databases", No. 4, ACM SIGMOD RECORD, December 1994, pp. 27-33.

[124] Y. Kiyoki, T. Kitagawa and T. Hayama, "A Meta-database System for Semantic Image Search by a Mathematical Model of Meaning", No. 4, ACM SIGMOD RECORD, December 1994, pp. 34-41.

[125] H.T. Anderson and M. Stonebraker, "SEQUOIA 2000 Metadata Schema for Satellite Images", No. 4, ACM SIGMOD RECORD, December 1994, pp. 42-48.

[126] W.I. Grosky, F. Fotouhi, I.K. Sethi and B. Capatina, "Using Metadata for the Intelligent Browsing of Structured Media Objects", No. 4, ACM SIGMOD RECORD, December 1994, pp. 49-56.

[127] U. Glavitsch, P. Schäuble and M. Wechsler, "Metadata for Integrating Speech Documents in a Text Retrieval System", No. 4, ACM SIGMOD RECORD, December 1994, pp. 57-63.

[128] F. Chen, M. Hearst, J. Kupiec, J. Pederson and L. Wilcox, "Meta-data for Mixed-media Access", No. 4, ACM SIGMOD RECORD, December 1994, pp. 64-71.

[129] M.A. Hearst, "Multi-paragraph Segmentation of Expository Text", Proc. 32nd Annual Meeting of the Association for Computational Linguistics, Las Cruces, New Mexico, 1994.

[130] V. Gudivada, "$\theta\Re$-String A Geometry-based Representation for Efficient and Effective Retrieval of Images by Spatial Similarity", Technical Report, TR-19444, Department of Computer Science, Ohio State University, Athens, 1994.

[131] S. Haykin, *Neural Networks : A Comprehensive Foundation*, Macmillan College Publishing Company, New York, 1994.

[132] R. Ng and J. Yang, "Maximizing Buffer and Disk Utilization for News On-Demand", Proceedings of Very Large Databases, 19994.

[133] H. Zhang, C.Y. Low and S.W. Smoliar, "Video Parsing and Browsing Using Compressed Data", Multimedia Tools and Applications, Vol. 1, No. 1, 1995, pp. 89-112.

[134] J.K. Wu, A.D. Narasimhalu, B.M. Mehtre, C.P. Lam, and Y.J. Gao "CORE : Content-Based Retrieval Engine for Multimedia Information Systems", Multimedia Systems, Springer-Verlag, 3(1), Feb. 1995, pp. 25-41.

[135] Dong-Yong Oh, Arun Katkare, Srihari Sampathkumar, P. Venkat Rangan and Ramesh Jain, "Content-based Inter-media Synchronization", Proc. of SPIE'95, High Speed Networking and Multimedia Systems II Conference, San Jose, CA, February 1995.

[136] V. Gudivada and V. Raghavan, "Design and Evaluation of Algorithms for Image Retrieval by Spatial Similarity", ACM Transactions on Information Systems, April 1995.

[137] S. Berson, L. Golubchik and R. Muntz, "Fault Tolerant Design of Multimedia Servers", Proceedings of ACM SIGMOD '95, San Jose, 1995 pp. 364-375.

[138] G.P. Babu, B.M. Mehtre and M.S. Kankanhalli, "Color Indexing for Efficient Image Retrieval", Multimedia Tools and Applications, Vol. 1, No. 4, November 1995, pp. 327-348.

[139] S.Adali, K.S. Candan, S.S. Chen, K. Erol and V.S. Subrahmanian, "Advanced Video Information System : Data Structures and Query Processing", Proceedings of First International Workshop on Multimedia information Systems, Washington D.C., September, 1995. Also, to appear in ACM/Springer Multimedia Systems.

[140] D.J. Gemmel, H.M. Vin, P. Venkat Rangan, L.A. Rowe, "Multimedia Storage Servers : A Tutorial", IEEE Computer, 1995, pp. 40-49.

[141] M.T. Ozsu, D. Szafron, G. El-Medani and C. Vittal, "An Object-Oriented Multimedia Database System for a News-on-Demand Application", ACM/Springer-Verlag Multimedia Systems, 1995.

[142] H. Blanken, "SQL3 and Multimedia Applications", Advanced Course on Multimedia Databases in Perspective, University of Twente, The Netherlands, 1995, pp. 155-178.

[143] P.A.C. Verkoulen and H.M. Blanken, "SGML/HyTime for Supporting Cooperative Authoring of Multimedia Applications", Advanced Course on Multimedia Databases in Perspective, University of Twente, The Netherlands, 1995, pp. 179-212.

[144] A. Analyti and S. Christodoulakis, "Multimedia Object Modeling and Content-Based Querying", Advanced Course on Multimedia Databases in Perspective, University of Twente, The Netherlands, 1995, pp. 213-238.

[145] C. Faloutsos, "Indexing Multimedia Databases", Advanced Course on Multimedia Databases in Perspective, University of Twente, The Netherlands, 1995, pp. 239-278.

[146] E. Bertino, B. Catania and E. Ferrari, "Research Issues in Multimedia Query Processing", Advanced Course on Multimedia Databases in Perspective, University of Twente, The Netherlands, 1995, pp. 279-314.

[147] P.M.G. Apers, "Search Support in a Distributed, Hypermedia Information Systems", Advanced Course on Multimedia Databases in Perspective, University of Twente, The Netherlands, 1995, pp. 315-336.

[148] S. Marcus and V.S. Subrahmanian, "Towards a Theory of Multimedia Database Systems", Multimedia Database Systems : Issues and Research Directions, Eds. V.S. Subrahmanian and S. Jajodia, Springer-Verlag, 1995, pp. 6-40.

[149] V.N. Gudivada, V.N. Raghavan and K. Vanapipat, "A Unified Approach to Data Modeling and Retrieval for a Class of Image Database Applications", In Multimedia Database Systems : Issues and Research Directions, Eds. V.S. Subrahmanian and S. Jajodia, Springer-Verlag, 1995, pp. 41-82.

[150] M. Arya, W. Cody, C. Faloutsos, J. Richardson and A. Toga, "Design and Implementation of QBSIM, a 3D Medical Image Database Systems", In Multimedia Database Systems : Issues and Research Directions, Eds. V.S. Subrahmanian and S. Jajodia, Springer-Verlag, 1995, pp. 83-104.

[151] A.P. Sistla and C. Yu, "Retrieval of Pictures Using Approximate Matching", In Multimedia Database Systems : Issues and Research Directions, Eds. V.S. Subrahmanian and S. Jajodia, Springer-Verlag, 1995, pp. 105-116.

[152] H.V. Jagdish, "Indexing for Retrieval by Similarity", In Multimedia Database Systems : Issues and Research Directions, Eds. V.S. Subrahmanian and S. Jajodia, Springer-Verlag, 1995, pp. 168-187.

[153] A. Belussi, E. Bertino, A. Biavasco and S. Risso, "A Data Access Structure for Filtering Distance Queries in Image Retrieval", In Multimedia Database Systems : Issues and Research Directions, Eds. V.S. Subrahmanian and S. Jajodia, Springer-Verlag, 1995, pp. 188-216.

[154] Banu Özden, R. Rastogi and Avi Silberschatz, "The Storage and Retrieval of Continuous Media Data", In Multimedia Database Systems : Issues and Research Directions, Eds. V.S. Subrahmanian and S. Jajodia, Springer-Verlag, 1995, pp. 240-264.

[155] S. Marcus, "Querying Multimedia Databases in SQL", In Multimedia Database Systems : Issues and Research Directions, Eds. V.S. Subrahmanian and S. Jajodia, Springer-Verlag, 1995, pp. 265-279.

[156] R. Cutler and K.S. Candan, "Multimedia Authoring Systems", In Multimedia Database Systems : Issues and Research Directions, Eds. V.S. Subrahmanian and S. Jajodia, Springer-Verlag, 1995, pp. 280-297.

[157] V. Kashyap, K. Shah and A. Sheth, "Metadata for Building Multimedia Patch Quilt", In Multimedia Database Systems : Issues and Research Directions, Eds. V.S. Subrahmanian and S. Jajodia, Springer-Verlag, 1995, pp. 298-318.

[158] A. Hampapur, *Designing Video Data Management*, PhD Thesis, Department of Computer Science and Engineering, University of Michigan, 1995.

[159] S.V. Raghavan, B. Prabhakaran and S.K. Tripathi, "Synchronization Representation and Traffic Source Modeling in Orchestrated Presentation", IEEE Journal on Selected Areas in Communication, special issue on Multimedia Synchronization, Vol. 14, No. 1, January 1996, pp. 104-113.

[160] J. Schnepf, J.A. Konstan and D.H.-C. Du, "Doing FLIPS: Flexible Interactive Presentation Synchronization", IEEE Journal on Selected Areas in Communications, Vol. 14, No. 1, January 1996.

[161] Banu Özden, R. Rastogi and Avi Silberschatz, "On the Design of a Low-cost Video-on-Demand Storage System", ACM/Springer Multimedia Systems, Vol. 4, No. 1, 1996, pp. 40-54.

[162] K. Selçuk Candan, V.S. Subrahmanian, and P. Venkat Rangan, "Towards a Theory of Collaborative Multimedia", IEEE International Conference on Multimedia Computing and Systems, Hiroshima, Japan, 1996.

[163] K. Selçuk Candan, B. Prabhakaran, and V.S. Subrahmanian, "Collaborative Multimedia Documents: Authoring and Presentation", Technical Report: CS-TR-3596, UMIACS-TR-96-9, University of Maryland, College Park, Computer Science Technical Report Series, January 1996.

[164] K. Selçuk Candan, B. Prabhakaran, and V.S. Subrahmanian, "Retrieval Schedules Based on Resource Availability and Flexible Presentation Specifications", Technical Report: CS-TR-3616, UMIACS-TR-96-21, University of Maryland, College Park, Computer Science Technical Report Series, 1996.

[165] S.V. Raghavan, B. Prabhakaran and S.K. Tripathi, "Handling QoS Negotiations In Orchestrated Multimedia Presentation", to be published in the journal of High Speed Networking.

[166] V. Balasubramanian, "State of the Art Review on Hypermedia Issues and Applications", http://www.csi.ottawa.ca/ dduchier/ misc/hypertext_review/

[167] P.M.E. de Bra, "Hypermedia, Structures and Systems", http://www.win.tue.nl/win/cs/is/debra/cursus/

[168] "CERN, presentation on World-Wide Web", http:/info.cern.ch/ hypertext/WWW/Talks/General.html

[169] Multimedia Toolbook 3.0 Users Guide, Asymetrix.

[170] IconAuthor 6.0 User's Guide, AimTech.

[171] Director 4.0 User's Guide, Macromedia.

GLOSSARY

Audio Data : Digitized representation of audio signals. Interpretation of audio data is based on its relationship to a constantly progressing time scale. Volume of the audio data depends on the required *quality, e.g.,* voice quality audio results in 64 Kb/s and CD quality audio results in 1.4 Mb/s.

B-Trees : An *n*-ary branched balanced tree.

Cluster Generation : Objects with similar features are grouped to form a cluster.

Content Based Querying : Queries in multimedia database management systems referring to the content of a stored object. (e.g.) *Show me the image of the person who has green hair.*

Continuous Media : Characterized by large amounts of data, high data rates, and temporal requirements. (e.g.) digital video and audio.

Delay : Maximum delay that might be suffered by a data unit during its transmission through the computer network. Expressed in terms of an absolute or a probabilistic bound.

Delay Jitter : Delay variation in data transmission.

Deterministic QoS Guarantee : Network service provider offers a strict commitment to guarantee the QoS agreed upon.

Discrete Media : Characterized by lack of temporal requirements. (e.g. text, graphics and images).

Disk Striping : Object is decomposed into a specified number of subobjects, denoted as *striping units*, which are allocated across different disks.

Flexible Temporal Specification : Providing a range of values for the time instants and durations of objects presentations.

Graphics Data : Representation of data such as images and drawings that can be generated based on formal descriptions, programs, or data structures.

Generated Data : Represents computer generated information such as animations and music.

Hard Temporal Specification : Providing deterministic values for the time instants and durations of objects presentations.

Hidden Markov Model : Has an underlying stochastic finite state machines that are defined by a set of states, an output alphabet, and a set of transition and output probabilities.

Histograms (color, graylevel): With values indicating the percentage of pixels that are most similar to the particular color or graylevels.

Inverted Files : A data structure that maps a word or an atomic search item, to the set of documents, or set of indexed units, that contain the word, its *postings*. A posting may be a binary indication of the presence of that word in a document or may contain additional information such as its frequency of occurrence in the document and an offset for each occurrence [45].

Metadata : Derived data describing the contents, structure, semantics, etc. of multimedia data. (e.g., data structure describing the objects contained in a video frame).

MIDI (Musical Instrument Digital Interface) : A detailed specification of a command set by which any complex musical score can be described as a series of commands. The command set includes note on/off for specific instruments and controller commands to control pitch blend, volume, reverberation effect, etc.

Mirroring : Replicating the contents of a disk to improve fault tolerance.

Multimedia Access : Appropriate access structures such as indexes and hash functions to aid in efficient retrieval of multimedia data.

Multimedia Storage : Storing of multimedia data that allows both reduction of the required amount of space and optimal retrieval of data.

Multiple Inheritance : Inheriting variables and methods from multiple superclasses.

Network Striping : Distributing the subobjects of an object across disks connected by a computer network.

Packet/Cell Loss : Probabilistic bound on the loss of a transmitted data unit, packet or cell.

Parity Scheme : Bit-wise exclusive-OR scheme to incorporate parity information in data storage. Used to improve fault tolerance.

Pictures/Image data : Digitized representation of drawings, paintings, photographs and prints.

Quality of Service (QoS) parameters : Guaranteed by the computer network service provider for distributed multimedia applications for transferring various media objects. Parameters include end-to-end throughput, delay, delay jitter, packet loss probability and inter-media synchronization requirements.

Query Predicates : The conditions that have to be satisfied for a data item to be selected as output data.

Query-by Example : The data item to be selected as output should be similar to the one presented in the example.

R-tree : Extension of the B-tree for multidimensional objects, with a geometric object being represented by its minimum bounding rectangle.

Semantic Information : Represents the meaning and use of data, with emphasis on the issues of contexts, ontologies, and their mappings to more representational issues (features, modalities, etc.) [121].

Segment Trees : Intervals that span lower level nodes may be stored in the higher level nodes of the index tree. Segment trees provide efficient mechanisms to index both *interval* and *point* data in a single index.

Shot : A shot is an unbroken sequence of frames from one camera. Defined by a beginning and an ending frame.

Speech Data : Represents spoken language (often not *defined* as an independent data type). Importance of speech is with respect to its use as an input/output mechanism for multimedia applications. Natural language processing techniques are needed to allow recognition of keywords and identification of specific speakers.

SQL3 : Enhanced version of SQL with new built-in data types such as Binary Large Objects (BLOB), new type constructors, and object oriented features.

SQL/MM : Enhanced version of SQL to support multimedia database applications. Still in the preliminary stages.

Staggered Striping : First fragment of consecutive subobjects are located at a distance of k disks within a cluster, where k is termed the *stride*.

Statistical QoS Guarantee : QoS guarantees may be met with a certain probability. Guarantees are given based on assumptions about the properties of particular types of media streams, not based on the maximum load.

Synchronization : Task of coordinating the presentation of multimedia objects in the time domain. In multimedia databases, the synchronization of object presentations are explicitly formulated.

Text Data : Often used to represent strings of characters. A complete definition of text includes structural information of documents such as title, authors as well as the layout information.

Throughput : Amount of data that will be sent through the network per unit time.

Video Data : Represents time-dependent sequencing of digitized pictures or images, called *video frames.* Time scale associated with video specifies the interpretation of each frame in absolute time. Regular motion video requires 25 or 30 frames/second (depending on the video standard employed).

ACRONYMS

ADT	Abstract Data Type
ATM	Asynchronous Transfer Mode
BLOB	Binary Large Objects
DCT	Discrete Cosine Transform
DDL	Data Definition Language
DFT	Discrete Fourier Transform
DML	Data Manipulation Language
DSSSL	Document Style Semantics and Specification Language
DTD	Document Type Definition
DTPN	Dynamic Timed Petri Nets
DVI	Digital Video Interactive
FFT	Fast Fourier Transform
FDDI	Fiber Distributed Data Interface
GIF	Graphics Interchange Format
HDTV	High Definition Television
HyTime	Hypermedia/Time-based Structuring Language
HMM	Hidden Markov Model

JPEG	Joint Photographers Encoding Group
LFS	Log-structured File Systems
MBR	Minimum Bounding Rectangle
MIDI	Musical Instrument Digital Interface
MHEG	Multimedia Hypermedia Experts Group
MPEG	Motion Pictures Encoding Group
NTSC	National Television System Committee
ODA	Office Document Architecture
OCPN	Object Composition Petri Nets
PAL	Phase Alternation Line
QBE	Query By Example
QoS	Quality of Service
RAID	Redundant Array of Inexpensive Disks
RSVP	Resource reSerVation Protocol
SAM	Spatial Access Method
SGML	Standard Generalized Markup Language
TIFF	Tag Image Format

INDEX